International Comparative Social Studies

VOLUME 32

The titles published in this series are listed at *brill.com/icss*

Oil and Security Policies

Oil and Security Policies

Saudi Arabia, 1950–2012

By

Islam Y. Qasem

BRILL

LEIDEN | BOSTON

Cover illustration: © Ruletkka|Dreamstime.com.

Library of Congress Cataloging-in-Publication Data

Qasem, Islam Y., author.
Oil and security policies : Saudi Arabia, 1950-2012 / by Islam Y. Qasem.
pages cm. -- (International comparative social studies, ISSN 1568-4474 ; volume 32)
Includes bibliographical references and index.
ISBN 978-90-04-27774-8 (hardback : alk. paper) 1. Petroleum industry and trade--Political aspects--Saudi Arabia. 2. Energy consumption--Political aspects--Saudi Arabia. 3. Internal security--Saudi Arabia. 4. National security--Saudi Arabia. 5. Security, International--Saudi Arabia. 6. Saudi Arabia--Foreign relations. I. Title.

HD9576.S32Q26 2015
338.2'72820953809045--dc23

2015028649

This publication has been typeset in the multilingual "Brill" typeface. With over 5,100 characters covering Latin, IPA, Greek, and Cyrillic, this typeface is especially suitable for use in the humanities. For more information, please see www.brill.com/brill-typeface.

ISSN 1568-4474
ISBN 978-90-04-27774-8 (hardback)
ISBN 978-90-04-27773-1 (e-book)

To my mother and the memory of my father

Contents

Acknowledgments

Most of the research for this book is derived from a PhD dissertation written at the Political Science Institute at Leiden University. I am truly grateful to a number of friends and colleagues, who supported, motivated, and guided me throughout this journey. I would like to acknowledge Professor Lisa Anderson for her advice and support during the research and writing of the dissertation. My dissertation promoter, Prof. Dr. Madeleine Hosli, at Leiden University kindly and patiently helped me tackle many challenges in the way of completion of the dissertation. Special thanks to the faculty of Political and Social Sciences at Universitat Pompeu Fabra (UPF), where I completed a Diploma of Advanced Studies. Their support and motivation encouraged me to study oil dependence and security. The Generalitat de Catalunya provided funding towards my PhD research, for which I am truly grateful. I am also deeply indebted to several anonymous readers for their helpful comments and criticisms.

Writing this book has been a joy and a challenge, but it could not have been accomplished without the love and support of my wife, Nynke. She not only tolerated my long working hours but also thoughtfully read and commented on previous and current drafts. My two little children, Noura and Kamil, brought infinite joy into my life while working on this book. Their cheerful presence gave me the necessary boost to write and rewrite. Finally, thanks to my parents, Irfat and Yasin Qasem, to whom this book is dedicated and who dedicated their lives to securing a better future for their children.

Islam Y. Qasem
The Hague
2014

List of Tables and Figures

Tables

Figures

List of Abbreviations

AOI	Arab Organization Industrialization
ARAMCO	Arabian American Oil Company
AWCS	Airborne warning and control system
b/d	barrels per day
BP	British Petroleum
EIA	Energy Information Administration
GDP	Gross Domestic Product
IMF	International Monetary Fund
IPC	Iraq Petroleum Company
Mbd	Million barrels per day
MDAA	Mutual Defense Assistance Agreement
MEES	Middle East Economic Survey
NATO	North Atlantic Treaty Organization
OAPEC	Arab Petroleum Exporting Countries. Members: Algeria (1970), Bahrain (1970), Kuwait (1968), Libya (1968), Syria (1972), UAE (1970), Egypt (1973), Iraq (1972), Qatar (1970), Saudi Arabia (1968), Tunis (1982)
OECD	Organization for Economic Cooperation and Development
OPEC	Organization of the Petroleum Exporting Countries. Members: Algeria (1969), Indonesia (1962), the Islamic Republic of Iran (1960), Iraq (1960), Libya (1962), Nigeria (1971), Qatar (1961), Saudi Arabia (1960), United Arab Emirates (1967), Venezuela (1960)
PLO	Palestinian Liberation Organization
SABIC	Saudi Arabian Basic Industries Corporation
SAMA	Saudi Arabian Monetary Agency
SOCAL	Standard Oil Company of California
SR	Saudi Riyal
UN	United Nations
USMTIM	United States Military Training Mission

Introduction

Oil dependence has fundamentally changed the conditions of world politics.[1] Not only has oil wealth played an essential role in domestic politics of producers, but also producers and consumers have exploited oil dependence as a political weapon in international relations. While producers have resorted to cutting production or allowing prices to surge, consumers have deployed oil sanctions against producers. Central to oil politics have been oil rentier states, oil-producing countries whose economic and political survival heavily relies on oil resources.[2]

This book gives an insightful account of the ways in which oil dependence has affected the internal and external security strategies of Saudi Arabia – an oil rentier state par excellence. As the world's most important oil producer and exporter, Saudi Arabia, has been the linchpin of the international oil market and the bulwark of political stability in the Middle East for more than four decades. It has been able to exert unparalleled influence on prices and supplies, with far-reaching political and economic consequences for consumers and other producers. At the same time, the boom-bust oil cycle has determined the wealth and power of the Saudi rentier state. Periodic shifts between boom and bust have resulted in different patterns of oil dependence, which, in turn, have influenced the Saudi state's ability to respond to internal and external threats.

Since its creation in 1932, Saudi Arabia has confronted a range of military and ideological threats. At various times, powerful countries in the region, including Gamal Abdul Nasser's Egypt, Saddam Hussein's Iraq and Ruhollah Khomeini's Iran, threatened the legitimacy of the Saudi ruling family and the territorial integrity of

---

1 Throughout the book, I use the terms "dependence" and "asymmetric interdependence" interchangeably. I also use the terms "interdependence" and "symmetric interdependence" interchangeably.

2 The concept of rentier state is fully explored in Chapter 1. In short, the definition of the rentier state is based on three conditions: first, state revenue is highly based on oil sales abroad or other foreign sources; second, public expenditures represent a significant share of GDP; and third, only a small fraction of the population is responsible for the generation of rent (state revenue). My definition is based on the following scholarly works: Hazem Beblawi and Giacomo Luciani, eds., *The Rentier State* (New York: Croom Helm, 1987); Jill Crystal, "Coalitions in Oil Monarchies: Kuwait and Qatar," *Comparative Politics* 21 (July 1989), 427–443; Ghassan Salame, "Political Power and the Saudi State," *MERIP Reports* (October 1980), 5–22; Jacques Delacroix, "The Distributive State in the World System," *Studies in Comparative International Development* 15 (1980), 3–21; and Lisa Anderson, *The State and Social Transformation in Tunisia and Libya, 1880–1980* (Princeton: Princeton University Press, 1986).

 | DOI 10.1163/9789004277731_002

the kingdom. No less threatening to the Saudi royal family has been the war of ideas in the Middle East. In the 1950s and 1960s, Nasser's ideology of pan-Arabism, which won hearts and minds throughout the Arab world, posed a serious internal and external threat to the kingdom. In the 1980s, the Saudi regime had to cope with ideological and military threats from Iran and their spillover to the Shiite minority in Saudi Arabia's Eastern Province. From the 1990s, militant Islamist ideology started to gain widespread appeal in the Muslim world, challenging the legitimacy of the ruling family and attacking the political status quo.

The nature and intensity of the Saudi state's responses to these and other threats varied in accordance with the different patterns of oil dependence. During the boom era, when global demand exceeded supply, Saudi Arabia had the opportunity to forge alliances with or to pressure consumers, and to vastly increase spending on welfare and security at home. In a glut market, soft demand and low prices translated into less revenue and power vis-à-vis consumers. However, with its vast reserves and export capacity, the Saudi state gained leverage over other producing countries by trying to manipulate oil prices through production adjustments.

Organization of the Book

The book is organized as follows: In Chapter 1, I introduce rentier theory and Saudi Arabia as a classical example of a rentier state. The chapter is in two sections. Section 1 includes a thorough discussion of the various dimensions of the rentier state, such as the rentier economy, the external source of state revenues and public expenditures. It concludes with a critique of rentier theory. In the second section, I provide a historical background of Saudi Arabia, describing the creation of the Saudi state and politics before the age of oil. Data on oil revenues and public expenditures are included to demonstrate the emergence of the Saudi rentier state.

In Chapter 2, I develop a model for analyzing the ways in which oil dependence affects the security behavior of rentier states. The chapter has five sections. Section 1 discusses the shortcomings of rentier theory and the two dominant international relations theories, realism and liberalism. Section 2 introduces neo-rentier theory, stressing the importance of opportunity costs for understanding dependence. Section 3 describes possible internal and external security strategies of rentier states. Section 4 explains the types of dependence between rentier states and consumers and producers, and how to assess opportunity costs. The last section includes casual mechanisms that link patterns of dependence to security strategies.

Chapters 3–6 analyze the effects of shifting patterns of oil dependence on Saudi Arabia's internal and external security strategies in the period 1950–2012. During this period, Saudi Arabia went through four stages of dependence: "pre-boom," "oil-boom," "oil-bust" and "oil-boom." The "pre-boom" era (1950–1970) was characterized by oil companies holding full control over all decisions related to oil exports and prices, and the Saudi state's dependence on royalties from oil companies. In the "oil-boom" period (1970–1985), crucial structural changes took place in the international oil market. Oil companies ceased to be responsible for the upward stream (the exploration and production of oil), and oil production and prices became the responsibility of oil-exporting countries. Moreover, the United States, a major oil-exporting country, transformed into a net importer. Consuming countries became dependent on producers in the course of this period. During the "oil-bust" period (1985–2000), worldwide demand for oil dropped significantly while excess supply increased sharply. As a result, producing countries became dependent on consumers. At the turn of the twenty-first century (ti), a new "oil-boom" era started as developing countries significantly increased their demand for oil.

The four chapters are organized similarly. Each one consists of three sections. Section 1 describes the pattern of dependence that existed at the time in question, focusing on world consumption and production as well as Saudi Arabia's export capacity and proven reserves. In Section 2, I consider the impact of the pattern of dependence on the intensity of internal validation and balancing. Section 3 discusses Saudi Arabia's responses to major threats.

Chapter 7 offers a conclusion by way of summarizing the insights from Chapters 4–7. Based on forecasted changes in the world's demand for and supply of oil, the chapter concludes with an analysis of Saudi Arabia's security behavior in the coming two decades.

CHAPTER 1

Rentier Theory and Saudi Arabia

This chapter provides background information on the concept of rentier theory and describes the rentier characteristics of the Saudi state. In the first section I review the primary literature on rentier theory and describe essential concepts, such as rentier economy and rentier state. Section 2 provides a historical background of Saudi Arabia. In this section, I describe the creation of the Saudi state as well as politics in the era before the discovery of oil. This section concludes with a description of Saudi Arabia as a rentier state.

What is a Rentier State?

The concept of the "rentier state" was first proposed by the economist Hossein Mahdavy to identify the effects of oil nationalization on the structure and source of economic growth in the Middle East, especially in the case of Iran after the mid-1950s. Mahdavy defined rentier states as "those countries that receive on a regular basis substantial amounts of external rent. External rents are in turn defined as rentals paid by foreign individuals, concerns, or governments to individuals, concerns or governments of a given country."[1] External rents are distinguished from foreign grants, because the latter are often of a temporary nature and uncertain. During the Cold War era, for example, several Middle Eastern states, notably Israel, Jordan and Egypt, received so-called "strategic rents" which are often listed as development aid in the World Bank statistics.[2] Although Madavy's definition of external rent could also include labor remittances, foreign direct investment, and other capital flows, his main concern is oil rent. Being mainly interested in the economic development of Middle Eastern countries before 1970, Mahdavy's examples of external rent include payments for the passage of ships through the Suez Canal, payments to oil transit countries, and oil revenue received by governments of exporting countries in the form of royalties from oil companies who were still in charge of oil exploration and development. In the early 1970s, however, nationalization of

1 Hossein Mahdavy, "The Pattern and Problems of Economic Development in Rentier States: The Case of Iran," in *Studies in the Economic History of the Middle East,* M.A. Cook, ed. (Oxford: Oxford University Press, 1970), 428.

2 Alan Richards and John Waterbury, *A Political Economy of the Middle East* (Boulder: Westview Press, 1996), 17.

 | DOI 10.1163/9789004277731_003

the oil industry in many oil-exporting countries modified the nature and amount of rent. Consequently, the nature of oil revenue was transformed from royalties and taxes to revenues from selling oil abroad directly.[3]

While all countries in the Middle East in the 1960s received external rent, the variation in added value it constituted to their GNP and the percentage it represented of their total government revenue varied significantly across countries. Therefore, "the stage at which a country can be called a rentier state is determined arbitrarily," and Mahdavy is mainly interested in those cases where "the effects of the oil sector are significant and yet the rest of the economy is not of secondary importance."[4] The crucial point for Mahdavy is that rapidly increasing oil revenues transform governments into decisive players in the economy. Without resorting to taxation the state relies on oil revenues to finance large public projects and programs. Such dramatic increases in government expenditures, Mahdavy reasoned, would stimulate production by increasing demand. Higher demand, however, does not necessarily stimulate the productive sector of the economy, because oil revenues are largely used to import consumption goods, which cannot be produced at home. Therefore, oil revenues end up stimulating the productive sectors of countries from where goods are imported, not the local economy. The end result is that domestic economic growth remains unimpressive. A corollary of the rentier state economy is that social and political relations are also distorted by government expenditures and the absence of taxation. While small segments of the society greatly prosper from oil revenues, the rest of the society dwells in a state of underdevelopment. This situation of inequality, according to Mahdavy, does not necessarily provoke great friction because the elites are exploiting natural resources, not the people.

The Rentier State as a Sub-system of the Rentier Economy

Hazim Beblawi and Giacomo Luciani refined the concept of the rentier state in the 1980s. In *The Rentier State* Beblawi and Luciani refine the concept of the rentier state in several ways. First, the concept of state is redefined. Unlike Mahdavy's definition of state, which seems synonymous with "country," Beblawi and Luciani define the nature of the state as "the combination of essential indicators describing the relationship between the state and the economy."[5] The

3 Mahdavy, "The Pattern and Problems of Economic Development in Rentier States: The Case of Iran," 428–429.

4 Ibid., 431.

5 Hazem Beblawi and Giacomo Luciani, "Introduction" in Hazem Beblawi and Giacomo Luciani, eds., *The Rentier State* (New York: Croom Helm, 1987), 4. For a critique of the rentier theory see Michael Herb, *All in the Family: Absolutism, Revolution, and Democracy in the Middle Eastern Monarchies* (New York: State University of New York Press, 1999).

term "state" is given a dual meaning. First, by state is meant "the apparatus or organization of government or power that exercises the monopoly of the legal use of violence."[6] Second, the state is seen as being "synonymous with that of society and indicates the overall social system subject to government or power."[7] The first definition of state as "government," however, is the dominant usage in Beblawi and Luciani's writings. Second, they offer a distinction between what is meant by a "rentier state" and "rentier economy." The definition of rentier state as a state that derives a significant portion of its income in the form of rent from abroad is rejected. Instead, Beblawi and Luciani prefer to view the rentier state from an economic prism, embedding their definition of the rentier state within a broader definition of the "rentier economy." Thus, they write: "an economy substantially supported by expenditure from the state, while the state itself is supported from rent accruing from abroad; or more generally an economy in which rent plays a major role. A rentier state is then a sub-system associated with a rentier economy."[8]

According to Beblawi, four conditions should be present in order to classify a state as rentier. First, the situation of rent dominates the economy. In actuality, as Beblawi points out, there is no such thing as a pure rentier economy. "Each and every economy has some elements of rent."[9] A rentier economy, therefore, is characterized by the existence of substantial rent. The exact level of rent, according to Beblawi, "is a matter for judgment."[10] Second, not only does the rent need to be predominant but also external. Beblawi states:

> The externality of the rent origin is crucial to the concept of a rentier economy. The existence of an internal rent, even substantial, is not sufficient to characterize a *rentier economy*, though it could indicate the existence of a strong *rentier class* or group.[11]

A rentier economy based on internal rent does not exist without an internal productive sector. As such, domestic rent is nothing more than transfer of domestic payments. A substantial external rent, on the other hand, can sustain the economy without a productive domestic sector.

Third, only a small segment of the society is involved in the creation of wealth, while the rest is engaged in the utilization and distribution of the wealth. The

6 Ibid.
7 Ibid.
8 Ibid., 11.
9 Ibid., 51.
10 Ibid.
11 Ibid.

division of roles between the few as wealth creators and the many as wealth consumers differentiates the rentier from an open economy. For Beblawi an open economy with high foreign trade is not a rentier economy, because the majority – not the few – is engaged in the creation of wealth.

Fourth, in a rentier economy, the government is the principle recipient of external rent, and therefore control of rent is in the hands of the few. Because a portion of the rent inevitably has to be distributed among the population, the state becomes the most important actor in the economy. Not only does it provide public goods and services, but the state also becomes a distributor of benefits and favors. In other words, a sub-system of the rentier economy emerges which Beblawi calls "the rentier state."[12] Based on the above analysis, Luciani defines rentier states as "all those states whose revenue derives predominantly (more than 40 percent) from oil or other foreign sources and whose expenditure is a substantial share of GDP."[13] Next, I will further clarify the two main characteristics of the rentier state: external rent and public expenditures.

The Source of State Revenue

While Beblawi draws significant attention to the rent feature of the rentier state, Luciani, on other hand, stresses the question of whether the origin of state revenue is domestic or foreign. The fiscal sociological approach adopted by Luciani leads him to focus on the origin of revenue and the amount of public expenditures. In contrast to the tax state or what Luciani calls "esoteric states," which depend on domestic revenue and taxation, rentier states or "exoteric states" accrue revenue directly from abroad. Oil export is one way of receiving revenues from abroad, but is not the only one. Non-tax revenue also includes foreign aid and transit fees.

In the case of oil-exporting states what matters is not only production, but also the amount of export. In some oil-producing countries huge domestic consumption diminishes the amount of export, despite the voluminous production. Consider for example the case of the United States, especially since the early 1970s. The average total oil production of the United States in 1973 reached almost 9.2 mbd, but only a small portion of 230 thousand barrels was exported. In other words only 0.25 percent of total production was destined for the outside market. In fact, the amount of imports reached 6.25 mbd.[14] In contrast, in

12 Hazem Beblawi, "The Rentier State in the Arab World," in Beblawi in Hazem Beblawi and Giacomo Luciani, eds., *The Rentier State* (New York: Croom Helm, 1987), 49–62.

13 Ibid., 70.

14 United States Energy Information Administration (EIA), *Annual Energy Review 2006*, report no. DOE/IEA-0384. Accessed 2008. http://www.eia.gov/totalenergy/data/annual/previous.cfm#2006.

1973 Saudi Arabia produced 7.59 mbd but consumed only 0.5 mbd in 1973.[15] The lion's share of its production was exported, providing the state with a significant amount of wealth. The accrued wealth in turn freed the Saudi state from constraints of domestic revenue and taxation. As Chaudhry writes,

> Most of the taxes on Saudis and fees on resident foreigners were withdrawn; foreign companies were given five-year tax holidays (and extensions granted thereafter); and personal income taxes on foreign workers were eliminated.[16]

In other words, the external origin of revenue substitutes for taxation and transforms the tax state into a non-tax state or an "exoteric state."[17]

Public Expenditure

According to Luciani, the primary concern for the rentier state is wealth allocation. In rentier states, public expenditure represents a significant share of GDP, entailing expenditures on various welfare programs such as health, education, and infrastructure. Besides providing favors and benefits, the government also assumes the role of an important and ultimate employer in the economy. Civil employment, in other words, is transformed into nothing less than another channel for the government to distribute wealth and benefits. As Ayubi put it, "The creation of jobs in the oil state becomes almost an objective of its own right, with little regards for what these recruits should (or can) do."[18] This novel role of the government as the central distributor of wealth also sharpens the difference between what Luciani calls the allocation and the production (tax) state.[19] Take, for example, the case of the Arab Gulf countries. Oil income in Kuwait accounted for 84 percent and 62 percent of total revenue in 1980 and 1982, respectively, while government expenditure reached 53 percent of GDP in 1982. Oil and foreign grants accounted for 90 percent of total Omani state revenue in 1982. The Omani government expenditures were 55.5 percent and 47.3 percent of GDP in 1978 and 1982, respectively.

---

15 OPEC, *Annual Statistical Bulletin 2004*.

16 Kiren Aziz Chaudhry, *The Price of Wealth: Economies and Institutions in the Middle East* (Ithaca: Cornell University Press, 1997), 144.

17 Luciani, "Allocation vs. Production States: A Theoretical Framework," in Beblawi and Luciani, ed., *The Rentier State* (New York: Croom Helm, 1987), 69–71.

18 Nazih Ayubi, "Arab Bureaucracies: Expanding Size, Changing Roles," in Adeed Dawisha and I. William Zartman, eds., *Beyond Coercion: The Durability of the Arab State* (New York: Croom Helm, 1988), 22.

19 Ibid., 63–82.

High public expenditures alone, however, are not enough to distinguish the rentier state from other states. All states, rentier and non-rentier, Luciani points out, perform to different degrees the task of wealth distribution. Consider, for example, the case of Norway. Although Norway is one of the largest net exporters and the Norwegian government owns a majority share of Statoil (Norway's largest oil company), Norway does not meet the criteria of a rentier state. Neither government expenditure as percentage of GDP nor oil-gas revenue as percentage of total revenues meets Luciani's definition of a rentier state. As table 1.1 shows, despite the oil price peak in the late 1970s and early 1980s, oil revenues did not reach the threshold of 40 percent of total revenue. Likewise, although Norway is considered an example of a generous welfare state, its government expenditures do not rival the expenditures of rentier states found in the Persian Gulf.

TABLE 1.1 *Norway oil revenues (percentage of total revenue) and government expenditures (percentage of GDP)*

Year	Oil revenues as percentage of government revenue[1]	Governments expenditure as percentage of GDP[2]
1977	8.32	22.38
1978	12.34	24.01
1979	17.89	24.44
1980	26.90	20.79
1981	27.99	21.22
1982	28.14	21.46
1983	30.37	21.25
1984	32.40	20.72
1985	30.69	21.93
1986	18.61	22.86
1987	16.89	24.62
1988	14.61	25.69
1989	21.30	25.75
1990	23.56	25.86
1991	25.19	26.26

SOURCE: (1) STATISTICS NORWAY (STATISTISK SENTRALBYRÅ), HISTORICAL STATISTICS 1994. (2) FOR GDP: UNITED NATIONS STATISTICS DIVISION, NATIONAL ACCOUNTS MAIN AGGREGATES DATABASE.

Saudi Arabia: Historical Background

Conquests and State-Making

The formation of the Saudi Arabian state followed a course of organized violence. Military organization and deployment served multiple and complementary purposes, including the protection of constituencies, obtaining of fresh sources of revenue, and elimination of rivals.[20] As early as 1900, motivated by the necessity of securing trading routes, Najd's merchants began to support Ibn Saud's military expansion.[21] Advancement of loans from the merchants helped facilitate Ibn Saud's war-making effort, which successfully checked nomadic raids but also eliminated the influence of his main rival Al Rashids in Najd. By 1906 the conquest of Najd was completed. Yet Ibn Saud devoted two more decades to war-making, expanding his military adventures throughout Arabia to redeem what, by his account, belonged to him and his household. Throughout these military adventures his economic position was precarious and his military capabilities were rudimentary.

The military capabilities improved with the creation of the *Ikhwān* in 1912 (discussed in more detail in the following section).[22] Before the *Ikhwān* was created Ibn Saud had no standing army; his regiment was a contingent of sedentary people (*hadar*) made up of villagers and merchants from Central Arabia. Although the villagers and townsmen were the most loyal to Ibn Saud, they were ill-suited for rapid mobilization and extended warfare activities. With the creation of the *Ikhwān*, however, Ibn Saud acquired not exactly a standing army, but a dependable force, highly mobilized, loyal, and courageous. They were audacious warriors empowered by a religious zeal impelling them to fight and die for the cause of God (*fi sabil Allāh*).[23] Thus, they were highly instrumental in the pursuit of Saudi territorial expansion: namely, the mountain area of Asir in 1920, Hail in 1921, Jauf in 1922, and lastly the Hijaz in 1925.

---

20 Charles Tilly, ed., *The Formation of National States in Western Europe* (Princeton: Princeton University Press, 1975).

21 Rayed Khalid Krimly, *The Political Economy of the Rentier States: A Case Study of Saudi Arabia (Ph.D. Dissertation: The George Washington University, 1993)*.

22 The meaning of the Arabic word *Ikhwān* is "brethren." It is often used in the context of brethren in religion.

23 For an excellent account of the *Ikhwān* see John S. Habib, *Ibn Saud's Warriors of Islam* (Leiden: E.J Brill, 1978). For an authoritative account of the relationship between Al Wahhab and Al Saud see Christine Moss Helms, *The Cohesion of Saudi Arabia: Evolution of Political Identity* (London: Croom Helm, 1981).

Under dire economic circumstances, Ibn Saud turned to the gains of war to make further war. The battles were financed by the booty taken in raids (*ghanima*), in addition to religious taxation (*zakat*). *Zakat* was calculated on the basis of 10 percent of mostly livestock, though as a source of revenue *zakat* was contingent on the rainfall. The arid climate of Central Arabia brought more drought than rain, and with less rain there was less pasturing and consequently less zakat; As a second source of revenue Ibn Saud was also dependent on *ghanima*. The conquest of al-Hasa or al-Ahsa illustrates the significance of *ghanima* as a source of revenue. Philby observed: "Both here [Qatif] and in the towns and villages of Al Ahsa he [Ibn Saud] was able to replenish his sadly depleted stores and money-bags."[24] Conquest, however, was not financially rewarding only because of *ghanima*; it also increased revenues from taxation.[25] *Zakat* and *ghanima* continued to be the primary sources of revenue until the annexation of the Hijaz in 1925.[26] With the conquest of Hijaz, Ibn Saud secured a steady source of revenue from pilgrims' fees.[27] In the period of 1926–1929, 130,000 pilgrims generated state revenue equivalent of £4 million to £5 million.[28] By 1932, as an outcome of three decades of "war making," Ibn Saud achieved the second part of Tilly's maxim, "state making." His laborious effort, however, was far from over, and the last two decades of his life were dedicated to preventing the dissolution of his dominion under pressure of economic hardship and external threats.

With the rise and fall of the Saudi state in earlier times, the lesson of history has shown that establishing a reign by no means ensures its survival.[29]

---

24 Harry St. John Philby, *S'audi Arabia* (New York: Arno Press, 1972), 268.

25 Philby adequately summarizes the impact of conquest on revenues: "The great Wahhabi lived to see the meager £50,000 of his first years at Riyadh (1902–1912) double itself to £100,000 with the conquest of al Ahsa (1913/25), and rise steeply to an average of about four to five millions after the occupation of the Hijaz (1926/37)." Philby, *S'audi Arabia*, 333.

26 British subsidies also contributed to the revenue of Ibn Saud.

27 Hijaz is the home to the holiest and second holiest places in Islam, Mecca and Medina. The pilgrimage to Mecca is one of the five pillars of Islam. Every able Muslim with financial means is obliged to perform the *Hajj* once in a lifetime. The other four pillars are (1) declaration of faith (*shahāda*), (2) five times daily prayer (*salāa*), (3) fasting the month of Ramadan (*sawm*), (4) alms (*zakat*), which is required once a year.

28 Before the fall of the Hijaz the main sources of revenue were subsidy from Britain (£5,000 per month), tithe, and the booty.

29 The First Saudi Realm (1744–1818) and the second Saudi Realm (1824–1891) were short-lived and confined to the region of Najd. The first Realm was defeated by the Egyptian army, while the second Saudi state was overthrown by the Al Rashid family of Hail.

His dominion was politically fragile and economically entrenched in chronic financial difficulties until the end of the Second World War. Nature was a key factor behind his economic troubles. Arid climate eroded the conditions for farming and cattle pasturing, with approximately only 0.2 percent of the total area of the country representing cultivated land.[30] Low precipitation jeopardized taxes on oasis farming and nomadic pastoralism, one of the main sources of state revenue.[31] To makes matters worse, not only were nomadic tribes an unstable source of revenue, but they were also a heavy financial liability, as state grants and subsidies were used to buy their loyalty and discourage tribal raiding.[32]

But nature wasn't entirely unkind to Ibn Saud. Notwithstanding the forbidding climate, there was plenty of oil. The signing of the oil concession between Ibn Saud and Standard Oil California (SOCAL) in 1933 unlocked the financial constraints and charted, from an economic viewpoint, a prosperous future that hardly seemed imaginable at that time. In the immediate term, especially after the Second World War, oil money armed Ibn Saud with the financial means to prevent the disparate parts of his kingdom from disintegration and to lubricate the administrative apparatus. Had oil not been discovered and royalties poured into Ibn Saud's pouch, it is unlikely that an intact dominion would have been bequeathed to his oldest son Saud. King Saud inherited a financially viable but minimal state, as its function hardly exceeded providing order and security, and its institutional structure was no more than the court of the king (*majlis*).

30 Cultivated area is only a little over 0.2 percent of the total area of the country, and about 85 percent of the cultivated area relies on irrigation. See Peter Beaumont, "Water and Development in Saudi Arabia," *The Geographical Journal*, no. 143 (March 1977): 42–60. See also Daniel Schofield, ed., *The Kingdom of Saudi Arabia* (London: Stacey International, 1986).

31 Ibn Saud himself put to Rihani: "When we get enough rain…the flocks are doubled and the people of Najd are happy. But in years of drought they are poor and wretched." See Ameen Rihani, *Ibn Sa'oud of Arabia: His People and His Land* (London: Constable & Co., 1928), 223.

32 The climate influenced the revenue, but also caused inter-tribal wars. In times of long drought, nomadic tribes were forced to leave the *Dirah* and head toward green pasture. Occasionally, however, when grassing is a matter of life or death, infringing on another *Dirah* became a source of conflict. *Dira* is an Arabic word for a delimited territorial domain under the control of a tribe, and where access to pasture and water in this domain is sanctioned to tribal members and protected against other tribes.

Domestic Politics in the Pre-Oil Era

By and large, Ibn Saud was an absolute monarch. His source of legitimacy was based on two modes of authority, tradition and charisma.[33] Ibn Saud possessed the leadership qualities characterizing what Max Weber identified as "charismatic" authority. Not only was Ibn Saud hailed from the well-known Anaza clan, but he also combined characteristics that are highly praised among tribesmen. These are piousness, tribalcraft,[34] and military prowess.[35] A demonstration of his leadership was staged at a conference of notables held in Riyadh in 1928 and attended by eight hundred delegates. Ibn Saud called the conference ostensibly to settle his dispute with *Ikhwān* chiefs, but the underlying goal was to secure tribal support and isolate his opponents. Ibn Saud raised the question of rulership, offering to abdicate if the delegates preferred to elect a new leader from the royal family. The crowd, however, rejected his resignation with cries that "we'll have no one else than you to rule us,"[36] reaffirming their allegiance.

Through a combination of charisma and institutionalization of Wahhabism, Ibn Saud held in his hand a monopoly over religion. As the most undisputed societal norm, Ibn Saud turned Wahhabism into a mechanism of coercion and mobilization.[37] In his own words, Ibn Saud claimed that

---

33 Weber defined charismatic authority in contrast to traditional as well as bureaucratic authority; however, in the case of Saudi Arabia these two concepts coexisted and reinforced each other, forming a cohesive authority structure. Max Weber, *The Theory of Social and Economic Organization*, trans., A.M. Henderson and Talcott Parsons (New York: The Free Press, 1964), 364.

34 By tribalcraft I mean having knowledge of tribal affairs and mastery of the art of dealing with them.

35 Many of the qualities possessed by Ibn Saud were also found in medieval kings; Bendix observed that " A ruler's function was at once military, political, and religious: He was expected to lead in war, settle disputes, and make sacrifices for victory, good crops, and peace." See Reinhard Bendix, *Kings or People: Power and the Mandate to Rule* (Berkeley: University of California Press, 1974), 25. For an examination of Ibn Saud leadership and qualities see Rihani, *Ibn Sa'ud of Arabia: His People and His Land*; Philby, *S'audi Arabia*; Philby, *Arabian Days: An Autobiography*; and Harry St. John Philby, *The Heart of Arabia: A Record of Travel and Exploration*, volume II (London: Constable & Company, 1922).

36 For a fuller discussion of this episode see Habib, *Ibn Sa'ud's Warriors of Islam: The Ikhwan of Najd and Their Role in the Creation of the Sa'udi Kingdom, 1910–1930*, 121–135.

37 Helm summarizes the importance of "Wahhabiism has clearly influenced all aspects of social, economic and political life in Saudi Arabia, particularly in Najd where it was conceived and nurtured." See Helms, *The Cohesion of Saudi Arabia: Evolution of Political Identity*, 78.

> Government...has been established in this wide desert, which with its power, brought all forces that are in it under its control and managed its administration by the virtue of the social teachings of the religion.[38]

More than just for coercion purpose religion was equally fundamental in legitimizing Ibn Saud's authority.[39] Religion provided Ibn Saud with the moral and legal ground to justify obedience and obligations from his society, in exchange for his adherence to the strict teachings and to defend the causes of Islam. Such a social contract espoused by the Wahhabi was based on the views of the thirteenth-century jurist of the Hanbali tradition, Ibn Taimiya. In line with the view of Ibn Taimiya, as Helms noted, "a government was considered Islamic by virtue of the support it gave Islam and to the *umma*; it was perfectly legitimate to accept the rule of anyone who followed the Sharia." Therefore, "The Wahhabis accepted the Al Saud as a legitimate and hereditary Islamic government and, notably, had referred to them as Imams ever since the mid-eighteenth century."[40] In this manner, as in medieval Christian kingship, religion played a key role in the establishment of an Islamic kingship in Saudi Arabia under the command of Ibn Saud.[41]

The tale of the *Ikhwān* exemplifies Ibn Saud's usage of religion as a force of legitimacy and mobilization. In order to proselytize the Bedouin into "a true believer" there was no force involved except the power of the Qur'anic text based on the exegesis of Abd al-Wahhab. To become "a true believer," one would have to abandon the nomadic lifestyle and settle in *Hijar* to pursue an ascetic life centered on the learning and observing of Wahhabi Islam.[42] With this power in his hand, Ibn Saud succeeded in disciplining the

---

38 Quoted in Helms, *The Cohesion of Saudi Arabia: Evolution of Political Identity*, 78.

39 Ibn Saud, of course, was not the first to use religion to justify his authority, either in Islam or other traditions. For instance, in the sixteenth century, the House of Hohenzollern used the Lutheran faith to justify their authority. See Bendix, *Kings or People: Power and the Mandate to Rule*, 157.

40 Helms, *The Cohesion of Saudi Arabia*, 81. *Umma* is the Arabic word for a universal Islamic community; Sharia is the Arabic word for Islamic law.

41 There are striking similarities between Lutheranism and Islam. One may even argue that Luther and Taimiya shared a similar view of state and religion. Both sought to maintain the supremacy of religion above the secular rulers, but had to compromise with kings who were willing to demonstrate reverence to religion, though they upheld political exigencies above all.

42 The process by which Ibn Saud re-Islamised the Bedouin was quite simple, according to Dickson, the British political officer at Kuwait in that time: "He [Ibn Saud] would send for the Shaikh and tell him in blunt terms that his tribe had no religion and that they were all

conflicting tribes into a standing and loyal religious militia called *Ikhwān*. Exploiting the *Ikhwān*'s enthusiasm for battling the "infidels,"[43] Ibn Saud expanded the frontier of his kingdom, enhancing the security of his emerging kingdom by eliminating the two primary rivals, the Rashids and Hashemites. First, with the help of the *Ikhwān,* Ibn Saud defeated the Al Rashids at Hail, employing religion as the principal motive behind his conquest. "He rallied all his wisdom in handling men and set out to convince the assembly that only by attacking the Rashids at Ha'il, could they serve their religion and country."[44] With the fall of the Rashids, Ibn Saud then directed the sword of the *Ikhwān* toward the conquest of Jauf in 1922 and, lastly, Hijaz in 1925. Behind the conquest of Hijaz were two overlapping incentives, the collection of pilgrimage tax and elimination of his second archrival, the Hashemite dynasty.

Although religion was highly effective as an instrument of mobilization, it was a means rather than an end. Ibn Saud himself had said:

> Politics and religion are not the same. But we the people of Najd desire naught that is not sanctioned by religion. Therefore, if religion sanctions our desire, the political measures we adopt for its realization must be lawful. If politics fail, then, war. And everything in war is permissible.[45]

Having indoctrinated the *Ikhwān* with a holy militancy, their allegiance served Ibn Saud's purpose in times of war; but the zeal of the *Ikhwān* did not differentiate between religion and politics, especially in times of peace. For the *Ikhwān* making war against the infidels at all times and places constituted the *sine qua non* of the "true believer." As the rift with the *Ikhwān* began to widen, Ibn Saud warned them, asking, "Are there not a number of you upon whose fathers' and grandfathers' necks my sword and that of my fathers and grandfathers

---

"Juhl" [meaning ignorance in Arabic – a term frequently used before the arrival of Islam]. He next ordered the Shaikh to attend the local school of 'ulama', which was attached to the great mosque of Riyadh, and there undergo a course of instruction in religion. At the same time half a dozen 'ulama', attended by some genuinely fanatical Akhwan, such as Al Duwaish the Shaikh of the Mutair were sent off to the tribe itself." Quoted in Habib, *Ibn Saud's Warriors of Islam*, 30.

43 Infidels are not only the non-Muslims, but also Muslims who did not follow Wahhabi orthodoxy, such as the Hashemite in the Hajz (Western Province) and the Shi`a in the Eastern Province.

44 Habib, *Ibn Saud's Warriors of Islam*, 85.

45 Quoted in Rihani, *Ibn Sa'ud of Arabia: His People and His Land*, 235.

made play?'"[46] The *Ikhwān*, however, hardened by the dogma of faith, would not retreat, and instead intensified their raiding against towns and tribes under the Ibn Saud's jurisdiction, thereby undermining his authority. Having failed to restrain the *Ikhwān*, Ibn Saud was caught between losing his throne, if not his life, and the destruction of the *Ikhwān*. Now that his holy warriors had turned against him, Ibn Saud turned to the townsmen for recruits. His charismatic authority allowed him to raise an army of 30,000 fighters to confront the *Ikhwān* in March 1929 in the battle of Sibila, where the rebels were defeated decisively.

Aside from security and order, Ibn Saud set no other societal goals for his state. The figures in Table 1.2 show a minimal state, lacking the apparatuses to penetrate and transform the society. Until 1950, there were four ministries: Foreign Affairs, Finance, Interior, and Defense.[47] The four ministries reflected concerns and objectives of the pre-oil state. Ministries of Finance and Foreign Affairs were to tackle the financial problems of the kingdom, while the latter two ministries were in charge of security and order. To these ministries was assigned a sizable portion of the budget; in combination with the budgets of tribal annuities and royal cabinet, the allocation reached 67 percent of total revenue and 103 percent of oil revenue. By contrast, the sum allocation to health, education, agriculture, and public work amounted to 22 percent of total revenue.

TABLE 1.2 *Saudi Arabia revenues (Million dollars), 1938–1944*

Year	Pilgrimage	Customs	Other local	Total except oil	Oil	Total
1938	2.63	3.67	0.53	6.83	0.34	7.17
1939	1.31	2.31	0.56	4.18	3.21	7.39
1940	1.31	2.31	0.62	4.24	4.79	9.03
1941	–	–	–	2.2	3.45	5.65
1942	–	–	–	1.9	3.41	5.31
1943	–	1.8	2.39	4.19	1.32	5.51
1944	3	1.5	3	7.5	1.66	9.16

SOURCE: YOUNG, SAUDI ARABIA: THE MAKING OF A FINANCIAL GIANT, 126.

46 Helm, *The Cohesion of Saudi Arabia: Evolution of Political Identity*, 114.

47 Ministry of Foreign Affairs was established in 1930, Finance in 1932, Interior in 1933, and Defense in 1946.

Financial Stress in the Early Years

The zenith of Ibn Saud's political realization, a royal dominion named after his family, however, was eclipsed by the shadow of a worldwide economic crisis. Not even the remote kingdom was able to escape the impact of the economic crisis. As agriculture prices slumped in places from where pilgrims came, such as India and the East Indies, the number of pilgrims to Mecca came to a halt.[48] It was under these dire economic straits that King Ibn Saud signed the historical concession with Standard Oil of California (SOCAL) in 1933, which ran for sixty years and covered a large area of the kingdom's eastern region. SOCAL consequently passed the concession to a wholly-owned subsidiary called California-Arabian Standard Oil Company (CASOC).[49] The commencement of the oil concession ushered in a new era, which altered all aspects of social, economic, and political life in Saudi Arabia. In exchange for the concession, SOCAL agreed to advance an immediate loan of 30,000 gold pounds, and eighteen months later a second loan of 20,000 gold pounds, plus an annual rental fee of 5,000 gold pounds.[50] Five years later, in 1938, when oil was discovered in large quantities, the Saudi government collected its first oil royalty of $340,000.[51] As shown in table two, in 1940 oil revenue reached $4.79 million, constituting roughly half of the total revenues. But the oil fortune lasted only until the outbreak of the Second World War. The war hostility curtailed not only oil production but also pilgrimage rent. The figures in table two show that total revenue dropped from $9 million in 1940 to $5.5 million in 1943.

As the two primary sources of revenue dwindled, the state once again plunged into a financial crisis that threatened to disintegrate the kingdom. To that effect, the American vice-consul in Yemen wrote to the Secretary of State: "Neither he [Ibn Saud] nor his government can be expected to last much longer without money."[52] By that time, however, the oil concession was of a significant value not only to the oil companies, but also to the United States government. Moreover,

48 According to Philby, Ibn Saud confided to him: "that if anyone were to offer me a million pounds now he would be welcome to all the concessions he wants in my country." Harry St. John Philby, *Arabian Days: An Autobiography* (London: Robert Hale Limited, 1948), 291.

49 The name was changed in 1944 from CASOC to Arabian-American Oil Company (ARAMCO).

50 Arthur N. Young, *Saudi Arabia: The Making of a Financial Giant* (New York: New York University Press, 1983), 8.

51 The concession set a fixed royalty to the government to about $1.65 a ton or 22 cents a barrel. See Seymour, *OPEC Instrument of Change*, (London: The Macmillian Press LTD, 1980), 13.

52 Quoted in Nadav Safran, *Saudi Arabia: The Ceaseless Quest for Security* (Ithaca: Cornell University Press, 1988), 58.

the prospect of discovering a large quantity of oil turned the desert kingdom into an invaluable prize that could be ignored neither by the British nor the Americans. In turn, the *raison d'état* of oil pitted old allies against one another in Arabia. In response to the king's plea for financial aid, both Britain and the United States devised various financial rescue schemes. Aid from Britain during the period of 1943–1944 reached $37 million.[53] On the other hand, in an effort to undermine the British from winning ground with Ibn Saud at the cost of American interest, in February 1943, President Roosevelt declared that Saudi Arabia was vital to the interest of the United States, and therefore authorized the lend-lease aid for Ibn Saud.[54] It was a turning point marking the decline of British influence and the rise of American supremacy in Arabia.[55] From the Saudi perspective, the arrival of American influence in the region opened up the possibility of improving both the security of the regime and territorial integrity.

With British and American financial aid, Ibn Saud managed to ride out the financial crisis. The end of the war facilitated an increase in oil revenues many thousand-fold. Oil production resumed, rising from 21 million barrels in 1945 to 2200 million barrels in 1949; correspondingly, Arabian American Oil Company (ARAMCO) royalties to the Saudi government increased from $4.3 million to $39.1 million.[56] The flow of oil in turn had two major security implications for the kingdom. First, oil revenues facilitated higher spending on

53 Young, *Saudi Arabia: The Making of a Financial Giant*, 12. According to Young, SOCAL advances to Ibn Saud reached $12 million by 1944. Ibid., 12. In addition to the aid, the oil companies, SOCAL and Texaco, were deeply concerned about British influence in Saudi Arabia; therefore they lobbied the United States Government to subsidy Ibn Saud in order to thwart the British ambitions. For its part, the British government considered Saudi Arabia under its sphere of influence in the Middle East. Therefore, the British were suspicious of the American oil companies and the ambitions of the United States government in the kingdom. The mistrust between the two countries reached the point where Churchill himself protested in writing to President Roosevelt "that the United States has a desire to deprive us of our oil assets in the Middle East on which among other things, the whole supply of our Navy depends." In response, Roosevelt wrote "that the British wish to horn in on Saudi Arabian oil reserves." See David Holden and Richard Johns, *The House of Saud* (London: Pan Books Ltd, 1982), 128–130.

54 See Irvine H. Anderson, *Aramco, the United States, and Saudi Arabia: a Study of the Dynamics of Foreign Oil Policy, 1933–1950* (Princeton: Princeton University Press, 1981). See Holden and Johns, *The House of Saud*, 128.

55 Since then Saudi-American relations have evolved over the years into complex multilevel relations. There is an extensive literature in the canon of regional studies, and less in international relations, which attempts to analyze this relationship.

56 Young, *Saudi Arabia: The Making of a Financial Giant*, 125–126.

security imperatives. As shown in Table 1.3, budget allocation to defense and security was very high, reflecting the centrality of security concerns. In addition to acquiring modern arms, a significant aspect of internal security was the preservation of tribal allegiance, which also received a significant portion of the security budget.[57] The actual Saudi defense and security establishment was estimated to be less than 20,000, including armed forces, royal body guards, police, and tribal reserves.[58] Besides the insignificant number of troops, for the most part, the arrival of modern equipment and training was gradual and slow. Taking this into consideration, it can be inferred that the amount of subsidies assigned to tribes appears to be the most dominant component of the budget. The sum of allocations of tribal annuities and security amounted to 50 percent of total revenues and 76 percent of oil revenue. Second, oil reserves turned the kingdom into an international asset worthy of protection and support. Although no official defense treaty was signed between the two countries, the United States gave security assurances that affirmed its commitment to the preservation of the kingdom's territorial integrity and sovereignty.

The Emergence of the Saudi Arabian Rentier State

As noted above, in May 1933 the first oil concession was signed between the American oil company SOCAL and the Saudi Arabian government. Despite his reluctance, king Ibn Saud authorized the concession because of financial hardship. The terms of the concession were favorable to both parties. Five years after the concession was signed, oil was discovered in Dammam in commercial quantities.[59] In 1939 King Ibn Saud turned the valve which delivered

57 In the 1920s the government pursued a policy of domesticating the tribes. Tribal lands and rights were abolished in 1925, and tribal chiefs were encouraged to settle in Riyadh and live on government grants and subsidies. Krimly, *The Political Economy of the Rentier States: A Case Study of Saudi Arabia*, 150.

58 This was the estimate given by the American officers in 1953. See Safran, *Saudi Arabia: The Ceaseless Quest for Security*, 68. Another estimate given by Krimly is 44,171 in the year 1937, including the 39,000 "tribal reserves." According to Krimly, tribal reserves were not included in the training and mobilization, and they were merely used as "paper-force" in order to allocate salaries and subsides to tribal masses. Krimly, *The Political Economy of the Rentier States: A Case Study of Saudi Arabia*, 152–153.

59 In 1936 SOCAL invited another American oil company, Texaco, to join the concession on the basis of 50 percent partnership. In return, Texaco provided SOCAL with financial support and additional distribution outlets. In 1948 two other American oil companies

TABLE 1.3 *Saudi Arabia budget allocation (Million SR), 1947–48*

Revenue & expenditure	1947–48
Total revenue	215
Oil revenue	141
Expenditures:	
Royal cabinet	17
Health directorate	5
Education directorate	7
Defence & security	76
Judiciary & ulama	2
Telegraph & post	4
Adminstration & finance	22
Tribes Annuities & salaries	31
Pilgrimage	5
Agriculture directorate	6
General debt	6
Transportation and public work	30
Region administration	4

SOURCE: MINISTRY OF ECONOMY AND PLANNING, ANNUAL STATISTICAL BOOK, 1965.

the first Saudi oil to the world, marking the birth of the Saudi oil rentier state. The flow of oil brought unprecedented cash to the newly born state. The Second World War disrupted the cash flow but was resumed thereafter on an even greater scale. Next, I consider the two main characteristics of the rentier state, the externality of state revenue and the centrality of government expenditures.[60]

joined the concession, Standard Oil Company of New Jersey (Exxon) and Socony-Vacum Oil (Mobil).

60 For a general background information on Saudi Arabia political economy see Tim Niblock, ed., *State, Society and Economy in Saudi Arabia* (London: Croom Helm Ltd, 1983); Daivd E. Long, *Kingdom of Saudi Arabia* (Florida: University Press of Florida, 1997); Tim Niblock and Monica Malik, *The Political Economy of Saudi Arabia* (London: Routledge, 2007).

Externality of State Revenue

Immediately after the Second World War ended, global demand for oil erupted at a pace in which supply exceeded demand. In the postwar recovery era European demand for oil surged, and much of the oil supplies had to be imported into European markets.[61] Middle Eastern oil, including Saudi Arabia, became the primary source of supply. In response to the high demand, ARAMCO increased production, especially after the completion of the Tapline oil pipeline, which runs from the Abqaiq oil fields in Saudi Arabia to the port of Sidon in Lebanon. Figures in table 1.4 show the steady increase in oil exports. The majority of Saudi oil was exported to three destinations: North America, Europe, and Asia and the Far East. In the early years, 1955–1986, Saudi oil exports to North America rarely exceeded 14 percent of total exports. After 1986, however, Saudi oil exports to North America as a percentage of total exports climbed and stayed somewhere in the range of 22–30 percent.[62] This increase was due to higher demand from the United States. The United States increased its imports from Saudi Arabia to 1 mbd, almost 47 percent over the previous years. In fact since 1988 the United States maintained this course of importing between 1 mbd to 2 mbd from Saudi Arabia.[63] Unlike North America, which became an important importer in the mid-1980s, Europe was an important destination of Saudi oil since the early 1950s. From 1955 to 1986, Europe received between 38 percent and 57 percent of total Saudi oil exports. After 1986, however, European imports from Saudi Arabia substantially declined (in contrast, the United States increased its oil imports from Saudi Arabia). Like Europe, Asia and the Far East were also important destinations of Saudi oil exports. From 1955 to 2000, Asia and the Far East received between 30 percent and 60 percent of total Saudi oil exports.[64]

Unlike the steady rise in oil exports, oil revenues show a greater degree of fluctuation because of various factors such as fluctuations in oil prices, changes in the terms of the agreement between ARAMCO and the Saudi

61 In the immediate years after the Second World War, due to extreme shortage of coal supplies, Europe experienced an energy crisis, which posed serious threats to the European recovery efforts. See Daniel Yergin, *The Prize: The Epic Quest for Oil, Money, and Power* (New York: Simon & Schuster), 422–424.

62 Saudi Arabian Monetary Agency (SAMA), *Annual Statistical Bulletin*, Various Years. Accessed 2008/2009. http://www.sama.gov.sa/sites/SAMAEN/Pages/Home.aspx.

63 See EIA, *Annual Energy Review*, 2007. Accessed October 2007. http://www.eia.doe.gov.

64 The percentages are based on data from various annual statistical books available from the Kingdom of Saudi Arabia, Ministry of Economy and Planning.

TABLE 1.4 *Saudi Arabia oil revenues as a percentage of total exports and total revenues, 1947–2012*

Year	Total exports (Million barrels)[1]	Oil exports as percentage of total exports[2]	Oil revenue as percentage of total revenues[3]
1947	–	–	66
1951	–	–	49
1952	–	–	–
1954	–	93	71
1956	360	100	–
1958	378	69	81
1960	471	77	79
1962	577	77	80
1964	653	78	84
1966	944	79	78
1968	1,123	104	76
1970	1,174	102	90
1972	1,993	115	88
1974	2,892	100	94
1976	2,940	101	89
1978	2,813	92	88
1980	3,376	93	92
1982	2,058	93	76
1984	1,168	97	71
1986	1,190	90	56
1988	1,245	83	57
1990	1,642	90	78
1992	2,409	93	76
1994	2,275	90	74
1996	2,236	89	76
1998	2,332	84	56
2000	2,282	92	83
2001	2,203	–	81
2002	1,929	–	81
2003	2,381	–	81
2004	2,487	–	81
2005	2,631	–	81
2006	2,566	–	81

Year	Total exports (Million barrels)[1]	Oil exports as percentage of total exports[2]	Oil revenue as percentage of total revenues[3]
2007	2,541	–	81
2008	2,672	–	81
2009	2,288	–	81
2010	2,425	–	81
2011	2,635	–	81
2012	2,784	–	81

SOURCES: COMPILED BY THE AUTHOR FROM (1) 1950–1969, MINISTRY OF ECONOMY AND PLANNING, ANNUAL STATISTICAL BOOK, 1965, 1970; 1970–2000, SAUDI ARABIAN MONETARY AGENCY (SAMA), STATISTICAL BULLETIN, VARIOUS YEARS. (2) INTERNATIONAL MONETARY FUND (IMF), INTERNATIONAL FINANCIAL STATISTICS YEARBOOK, VARIOUS YEARS. (3) FOR TOTAL REVENUES: SAME AS TOTAL EXPORTS.

Arabian government, and currency evaluation. Notwithstanding the swings in oil revenues, the centrality of oil for development and economic growth is evident in the substantial percentages relative to total exports and total revenues. Based on the domination of oil in exports, the economy of Saudi Arabia typifies what is known as a single commodity exporter. Such countries rely heavily on a single or few commodities for a substantial share of their revenues, and therefore they are vulnerable to price volatility. In the case of Saudi Arabia, the effect of price volatility is compounded by the so-called "resource curse."[65] Among the various deleterious effects of the resource curse is the failure to develop industries that produce goods for exports. The failure of Saudi Arabia to diversify away from oil can be seen by the fact that oil exports

65 The term "resource curse" is widely used to describe the negative effects of resource abundance on economic and political development. The term encompasses several mechanisms that show how the export of natural resources cause negative economic growth and distort domestic politics (for example, corruption, conflict, and undermining transition to democracy). The literature on the resource curse has been the subject of extensive research by economists and political scientists. See, for example, Jeffrey D. Sachs and Warner M. Andrew, "The Curse of Natural Resources," *European Economic Review*, no. 45 (2001): 827–838; Michael L. Ross, "Review: The Political Economy of the Resource Curse," *World Politics* 51 (January, 1999): 297–322. For recent work on this subject see Macartan Humphreys, Jeffrey D. Sachs, and Joseph E. Stiglitz, *Escaping the Resource Curse* (New York: Columbia University Press, 2007).

as a percentage of total exports did not decline, despite the ambitious development projects pursued by the state in the mid-1970s and thereafter.[66]

Centrality of Government Expenditures

As already indicated above, high government expenditure is considered to be the second important characteristic of the rentier state. The data in table five show the changes in the level of expenditure during different stages of oil extraction. In absolute terms the data indicate a steady increase in government expenditure starting in early 1960s, reaching a peak in the early 1980s before beginning to decline in the mid-1980. Starting in 1985, government expenditure experienced fluctuation due to changes in oil prices. As oil prices sunk, oil revenues drastically declined. For instance, as shown in Table 1.5, expenditure was halved in 1986, in comparison with 1981.

Close examination of expenditure trends raises several noteworthy issues. Although decline in oil revenues reduced the amount of expenditure relative to the boom period, it did not fundamentally alter the expenditure behavior of the state. Consider, for instance, that when oil prices bottomed in 1986 and 1988, the level of government expenditure remained high in absolute terms, consuming 43 percent of GDP. In order to accurately explain spending behavior, it is necessary to break government expenditure into its two parts, capital and current expenditure. Commentators on Saudi Arabia have often compared government expenditure of the boom and the bust in order to draw conclusions on domestic politics. The comparison, however, is misleading. In many of these studies government expenditure is treated as a whole – no efforts have been made to consider the composition of spending. To make an accurate assessment regarding the effects of expenditure it is necessary to compare the parts, and not just the whole. Without specifying the target category of expenditure, one may jump to quick conclusions – refuting rentier theory.

As shown in table five, except for certain years when spending was especially high, government expenditure during the so-called bust phase was not significantly different than in years of boom. The reason is that current expenditure remained significantly high even during the bust. When oil price reached $34 a barrel, 113 trillion Saudi Riyals (SR) was allocated to current expenditures. It dropped to 98 trillion SR in 1986 when oil prices sank to $13 a barrel. However, capital expenditure was drastically reduced. In 1981, capital expenditures reached 171 trillion SR, and by the year 1986 declined by 75 percent to only 38 trillion SR.

66 For a useful discussion on industrialization in the Middle East see Michel Chatelus and Yves Schemeil, "Towards a New Political Economy of State Industrialization in the Arab Middle East," *International Journal of Middle East Studies*, no. 16 (May 1984): 251–265.

TABLE 1.5 *Current and capital government expenditure as percentage of GDP, 1963–2012*

Year	Total expenditure[1]			GDP (Million SR[2])	Total expenditure as percentage of GDP
	Current expenditure	Capital expenditure	Total		
1963	–	–	2,686	8,670	31
1964	–	–	3,112	9,320	33
1965	–	–	3,961	10,400	38
1966	–	–	5,025	11,940	42
1967	–	–	4,937	13,140	38
1968	–	–	5,535	18,840	29
1969	3,853	2,175	6,028	20,190	30
1970	3,989	2,304	6,293	22,570	28
1971	4,756	3,374	8,130	30,500	27
1972	5,654	4,504	10,158	38,260	27
1973	8,470	10,125	18,595	53,530	35
1974	15,207	19,832	35,039	1,59,720	22
1975	37,931	43,304	81,235	1,63,670	50
1976	73,621	54,652	1,28,273	2,25,350	57
1977	71,417	66,631	1,38,048	2,60,960	53
1978	83,488	64,484	1,47,972	2,72,270	54
1979	1,02,447	83,277	1,85,724	3,75,470	49
1980	1,32,661	1,04,094	2,36,755	5,46,600	43
1981	1,13,636	1,71,014	2,84,650	6,22,180	46
1982	1,02,248	1,42,664	2,44,912	5,24,200	47
1983	1,24,052	1,06,134	2,30,186	4,45,210	52
1984	1,21,696	94,667	2,16,363	4,20,390	51
1985	1,19,865	64,139	1,84,004	3,76,320	49
1986	98,894	38,528	1,37,422	3,22,020	43
1987	1,34,419	50,500	1,84,919	3,20,930	58
1988	1,16,283	24,573	1,40,856	3,30,520	43
1989	1,18,303	36,567	1,54,870	3,57,060	43
1992	1,62,350	76,637	2,38,987	5,10,460	47
1993	1,84,878	3,012	1,87,890	4,94,910	38
1994	1,61,380	2,396	1,63,776	5,03,050	33
1995	1,48,776	25,167	1,73,943	5,33,500	33
1996	1,71,258	26,859	1,98,117	5,90,750	34

TABLE 1.5 *Current and capital government expenditure* (cont.)

Year	Total expenditure[1]			GDP (Million SR[2])	Total expenditure as percentage of GDP
	Current expenditure	Capital expenditure	Total		
1997	2,18,880	2,392	2,21,272	6,17,900	36
1998	1,71,163	18,897	1,90,060	5,46,650	35
1999	1,67,195	16,646	1,83,841	6,03,590	30
2000	2,16,958	18,364	2,35,322	7,06,660	33
2001	2,23,508	31,632	2,55,140	–	–
2002	2,03,500	30,000	2,33,500	–	–
2003	2,23,530	33,470	2,57,000	–	–
2004	2,47,649	37,551	2,85,200	–	–
2005	2,84,173	62,301	3,46,474	–	–
2006	3,22,411	70,911	3,93,322	–	–
2007	3,47,199	1,19,049	4,66,248	–	–
2008	3,88,839	1,31,230	5,20,069	–	–
2009	4,16,594	1,79,840	5,96,434	–	–
2010	4,55,043	1,98,842	6,53,885	–	–
2011	5,50,500	2,76,200	8,26,700	–	–
2012	6,11,626	2,61,679	8,73,305	–	–

SOURCES: COMPILED BY AUTHOR FROM (1) 1963–69, MINISTRY OF ECONOMY AND PLANNING,ANNUAL STATISTICAL BOOK, 1965, 1970; 197–2000, SAUDI ARABIAN MONETARY AGENCY, TABLES OF ANNUAL REPORT. (2) INTERNATIONAL MONETARY FUND (IMF), INTERNATIONAL FINANCIAL STATISTICS YEARBOOK, VARIOUS YEARS

Conclusion

This chapter introduced the concept of rentier theory. It detailed the two conditions that are essential for defining a rentier state: external source of revenue and allocation of wealth. Unlike the so-called productive state, which relies on taxation, rentier states directly derive the majority of revenues from abroad in the form of rent (payments from selling oil in the international market). Relieving the society from the burden of taxation, domestic politics of the rentier state, it is argued, can be summed up by the maxim "no representation without taxation." Not only does the state not fiscally coerce its population, it also engages in generous wealth distribution. As the second part of the definition stresses, the state provides a wide range of free services and subsidies. Although rentier theory has been a popular analytical tool in Middle East area studies, recent scholarship has questioned its explanatory power for interpreting domestic politics. Lacking sufficient sophistication for dealing with periods of boom and bust, critics have discredited the mechanism that wealth allocation ensures regime survival. In addition to its weakness in the domestic realm, the theory is mute on international politics – even though rentier theory is based on wealth transfer from the international economy. Notwithstanding its limited explanatory capability, the theory offers an underdeveloped but useful theoretical framework.

In the second part of the Chapter I provided a historical overview of Saudi Arabia. It outlined the process by which the Saudi state came to exist and described the political and financial situations before the arrival of oil wealth. The section concluded with an empirical description of the rentier nature of the Saudi state since the early 1950s. In terms of the magnitude of oil revenues and the portion of public spending as percentage of GDP, the data demonstrate that the Saudi state has been a rentier state par excellence.

CHAPTER 2

Neo-Rentier Theory

The proposed theoretical framework presented in this chapter aims to explain Saudi Arabia's domestic and international security behavior. The chapter is divided into five sections. The first section discusses the limitations of rentier theory as well as realism and liberalism in explaining rentier state's security strategies. Section 2 introduces the framework of neo-rentier theory, stressing the importance of opportunity cost for understanding relations of dependence. Section 3 describes internal and external security strategies of rentier states. Section 4 identifies different patterns of dependence and elaborates on measures of dependence between consumers and producers. Last section concludes with casual mechanisms linking the independent variable (patterns of dependence) to the dependent variable (security strategies).

Theoretical Limitations

Rentier Theory

Although proponents of rentier theory have attempted to interpret the domestic political dynamics of oil-producing countries, their analyses have been partially simplistic and ambiguous. In most cases the analysis of the rentier state did not go beyond conceptualization and descriptive analyses of the supposed correlation between rent and stability of an authoritarian regime (or lack of democratization). Indeed, rentier theory has a hard time explaining political implications under conditions of fluctuation in the oil market, which is not an unusual phenomenon in the daily life of the international political economy. In his analysis of the oil wealth effect on regime failure for 107 developing countries between 1960 and 1999, Smith concluded: "More importantly, despite what Beblawi and Luciani (1987) and others have argued, the durability effect appears to have been independent of consistent access to rents with which regimes can buy legitimacy, since the busts created no trend toward regime crisis or instability in exporting states."[1] Focusing on Saudi Arabia, Fandy observes that

> Had the assumption of the rentier model been sound, this period [referring to 1986 when oil price plunged] should have witnessed political turmoil.

1 Smith, "Oil Wealth and Regime Survival in the Developing World, 1960–1999," 242.

 | DOI 10.1163/9789004277731_004

> Because of the relative stability of the Saudi polity during that period and the rise of opposition at times of greater increase in oil prices, we must look beyond the rentier model when analyzing Saudi resistance.[2]

Not only does rentier theory not address effects of fluctuation in oil prices and levels of expenditures, but omits from the analysis other oil-based mechanisms. Besides these drawbacks, the theory is silent on rentier states' foreign policy. Although by definition the international economic system is vital for the survival of the rentier state, the theory does not address the role of oil in the relationship between domestic and foreign policies.

Realism and Liberalism

Realism has evolved and produced various theories; the most cited are classical realism, structural realism, and neoclassical realism. All three theories share common assumptions about world politics. These assumptions are the following: states are central units in the international system; states are rational actors; states are power-seeking as a means to an end; states live in an anarchic international system. Notwithstanding these similarities, each theory maintains a different research program, including divergent explanations and assumptions.

Classical realists claim that the source of power is political, economic, and military. In a rentier state, power is derived from providing other states access to oil resources and the ability to generate revenue from selling oil in the international economy. The logic of structural realism also fails to capture the idiosyncrasy of the rentier state, since the state is treated as a "black box." Without knowledge of the internal characteristics of the state, the concept of rentier economy, which is the most important aspect of the rentier state, disappears from the analysis. Neoclassical realism also fails to capture the economic logic of rentier state. For neoclassical realist, national power is determined by the ability to the state of extract resources from the society. In contrast, the primary source of power for rentier state is oil reserves.

The tradition of liberalism enjoys an equally important tradition of theoretical scholarship. Fundamentally, the liberal approach is based on specific ideals; these are individual freedom, international law, protection of human rights, and free trade. More than realism, the paradigm of liberalism is sensitive to the internal characteristics of the state and importance of economic interdependence. In the view of liberal scholars, international economic transactions

2 Fandy Mamoun, *Saudi Arabia and The Politics of Dissent*, (London: Macmillan Press Ltd, 1999), 26–29.

foster cooperation and peace among nations. John Stuart Mill writes: "It is commerce which is rapidly rendering war obsolete, by strengthening and multiplying the personal interests which are in natural opposition to it."[3] Similar views can be found in the works of Adam Smith, David Hume, and Joseph Schumpeter.[4] The positive correlation between commerce and peace remains a cornerstone in contemporary liberal political thought. However, as Baldwin has showed, "gain from trade" is another name for opportunity costs of breaking or altering the relations between two parties.[5]

Neo-Rentier Theory

Neo-rentier theory is based on the definition of the rentier state but modifies the logic of buying off legitimacy and incorporates assumptions and insights from liberalism and realism. It conforms to the realist logic that anarchy and relative military capabilities heightens the threat or use of military force and outbreak of wars.[6] In addition to the external military threat, the model assumes that a state strives to defend itself from political threat and domestically originated threats. Political threat aims to undermine the principles of legitimacy and the organizing ideology of the state or the regime. It can take various forms, including a propaganda assault, attempts to incite activities of rebellion, or providing support to dissident groups. Internal threat is credible for many developing countries. As Ayoob observes, "Threats to Third World states and regimes emanate from within their regions, if not from within these

3 John Stuart Mill, *Principles of Political Economy*, 4th reprint (Clifton: Kelley, 1973), 582.

4 Smith, *the Wealth of Nations*, Book III, Chapter IV; Joseph Schumpeter, *Imperialism and Social Classes*, trans., Heinz Norden (Oxford: Blackwell, 1951), 83–130.

5 Baldwin, "Interdependence and Power: A Conceptual Analysis."

6 For a discussion on the element of military threat, see Stephen M. Walt, *The Origins of Alliances* (Ithaca: Cornel University Press, 1990), 18–45. On the use of military force in international politics, see Robert J. Art and Kenneth N. Waltz, *The Use of Force: Military Power and International Politics* (Rowman & Littlefield Publishers, Inc., 2003); John J. Mearsheimer, *The Tragedy of Great Power Politics* (New York: W.W. Norton Company, 2001).

For a discussion on the element of military threat, see Walt, *Origins of Alliances*, 18–45.

For a fuller discussion on offensive and defensive implications see Robert Jervis, "Cooperation under the Security Dilemma," *World Politics*, no. 30 (January 1978): 167–214; Stephen W. Van Evera, *Causes of War: Power and the Roots of Conflict* (Ithaca: Cornell University Press, 1999). For a critique of the defensive/offensive impact on war see Jack S. Levy, "The Offensive/Defensive Balance of Military Technology: A Theoretical and Historical Analysis," *International Studies Quarterly*, no. 28 (January 1984): 219–238.

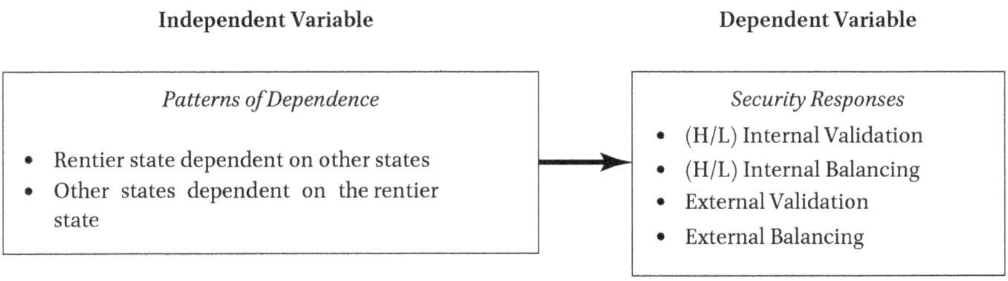

FIGURE 2.1 *Model of Neo-Rentier theory*

states themselves."[7] In other words, domestic and international threats are an important independent variable, which for the purpose of this study will be assumed to be constant.

Defined in terms of opportunity cost, rather than absolute or relative gain, patterns of oil dependence is treated the main independent variable. For country A to have influence over country B, the opportunity cost of breaking the relation must be high for B but not for A. In other words, if B has the ability to adjust with ease and speed to external pressure from A, then B might be sensitive but not vulnerable to A. Neither relative nor absolute gains are credible indicators of influence, however, since both obscure the fact that actors (for instance, B) might have alternative options that are not too costly. Therefore, relations of relative or absolute gains do not necessarily translate into relations of influence. Instead neo-rentier theory shows that different patterns of dependence influence the type and intensity of strategies used in the domestic and international arenas. These strategies are internal validation, internal balancing, external validation, and external balancing. Intensity indicates the level of spending on internal validation and internal balancing. The types of strategies available under different patterns of dependence are shown in Figure 2.1. The next two sections elaborate on the types of security strategies and patterns of dependence.

Dependent Variable: Security Strategies

In order to ensure regime stability and territorial integrity the state adopts various types of security strategies. The variation in the *type and intensity* of these

7 Mohammed Ayoob, "The Third World in the System of States: Acute Schizophrenia or Growing Pains?" *International Studies Quarterly*, no. 33 (March 1989): 70–71. Ayoob made a good case for why security in Third World countries is different than in Western countries. See Mohammed Ayoob, "Review: The Security Problematic of the Third World," *World Politics*, no. 43 (January 1991): 257–283.

strategies is the dependent variable. The literature of international relations describes six strategies that states consider in times of danger: internal balancing, external balancing, appeasement, buck-passing, bandwagoning, and blackmailing. These strategies, however, do not adequately capture the nature of the rentier state or account for its security goals. The main shortcoming is that all six strategies are externally oriented, while many rentier states suffer from internal and external vulnerabilities. Second, these strategies are based on the logic of relative military capability, and do not take into account economic interdependence. These shortcomings are rectified in two ways. First, the role of internal and external validation through economic incentive or coercion is considered as a security mechanism. Second, the strategies of internal and external balancing are reconceptualized to reflect the security role of economics and to account for internal threat. In what follows I provide a description of the security strategies discussed in the literature, but for the purpose of this study the emphasis is on internal and external validation as well as internal and external balancing.

Internal validation is the utilization of internal resources by the ruling regime in order to enhance its image at home and maintain support of the people. These mechanisms may take various forms, including appealing to shared societal values and norms, demonstrating charisma, adhering to a certain ideological line, or claiming divine or ancestral right to rule. Other attempts include the creation of public employment,[8] provision of free health and education, and extending generous subsidies. I use current and capital government expenditures to measure internal validation. External validation, on the other hand, according to Mastanduno et al., "refers to attempts by state officials to utilize their status as authoritative international representatives of the nation-state to enhance their domestic political position."[9] Indeed, this form of validation is not uncommon, especially in many Third World countries. For instance, many Arab leaders have assumed, at least in public, foreign policies that express an unyielding support for the Palestinian cause in order to increase their legitimacy at home. This type of validation can be accomplished in many ways, including supporting or rejecting United Nations resolutions, and breaking off diplomatic relations. To measure external validation I consider extension or suspension of financial aid, and manipulation of oil prices and supplies.

---

8 Consider for example, broadly speaking, that more than 50 percent of Saudi citizens in the year 1988 were employed in the public sector. See F. Gregory Gause III, *Oil Monarchies* (New York: Council on Foreign Relations Press, 1994), 59.

9 Mastanduno, Lake, and Ikenberry, "Toward a Realist Theory of State Action," *International Studies Quarterly*, no. 33 (December, 1989): 464.

Internal balancing, in the realist tradition, means that states rely on their own military capabilities in order to maximize security in the international system. Applying this definition to Saudi Arabia is problematic for two reasons: first, it does not take into account that the armed forces are mobilized not only for the purpose of external defense but also to check internal opposition;[10] second, it disregards the presence of a powerful network of internal repressive apparatuses concerned with internal security and the protection of the regime. The enormous spending on internal security apparatuses, such as the ministry of interior, security police, and domestic intelligence, reflects the graveness of the internal insecurity situation. Thus, in this study internal balancing is measured by spending on defense and internal security apparatuses.

External balancing, on the other hand, entails pooling resources and sharing the defense burden with other states (e.g., by forging alliances) for containing a dangerous state. Thus external balancing not only multiplies the number of guns and soldiers, but also is more economical than internal balancing. However, the virtue of alliance does not come without drawbacks. Alliances suffer from the difficulty of reaching an agreement among potential allies on a range of issues, such as coordination and distribution of burden, and perception of threats. Although realist literature has often portrayed external balancing in the form of pooling military resources, it is also equally valid to conceive balancing by means of pooling economic resources. In this book, oil policies that aim to benefit an ally, by raising oil output or moderating oil prices, are also considered an act of balancing. For instance, to support western powers Saudi Arabia may overproduce to make up for shortages in the oil market, or to mitigate oil prices.

Buck-passing is simply free-riding on someone else to counter an aggressor. It is an attractive option because a threatened state relies on the deterrence and military capabilities of another state, not its own, to check an aggressor. The most important disadvantage of buck-passing is the risk of failure on the part of the buck-catcher to successfully restrain the aggressor.[11]

10 For example the Saudi National Guard is a highly equipped and trained military force recruited along tribal lines for the purpose of defending the royal family from opposition rooted either in the society or the military. Gause, *Oil Monarchies*, 68.

11 According to Mearsheimer, there are four possible ways in which states take on the role of buck-passer. First, it seeks to improve its relations with the aggressor in the hope to shift the target of aggression to someone else, and hence rid itself of the burden of confronting the aggressor. Second, buck-passers suppress relations with the buck-catcher in order to avoid being dragged into confrontation or war. Third, the buck-passer builds up defensive capabilities in order to deter the aggressor or to be prepared for the possibility of confronting the aggressor. Fourth, the buck–passer allows, if not assists, the buck-catcher to

Blackmailing is the strategy least discussed in the international relations scholarship, but also is relevant to the rentier state. The *Oxford English Dictionary* offers a definition of blackmailing as an attempt to influence someone's behavior by using threats or manipulation. According to Mearsheimer, "*blackmail* is a more attractive alternative [to war], because it relies on the threat of force, not the actual use of force, to produce results."[12] Blackmailing need not, however, rely on the threat of force. Blackmailing can also be applied in the form of economic pressure, such as the threats of oil supply disruptions as ii the case of an embargo, or threats of blocking oil transit chokepoints such as the Strait of Hormuz and the Suez Canal.[13]

Alternatively, hoping to avoid or mitigate aggression, states pursue concessionary policies such as alliance dissolution, negotiation, or, most important, appeasement and bandwagoning. Unlike balancing, appeasement and bandwagoning are not intended to contain or confront an aggressor. Both strategies are adopted by a weak and isolated state when confronted by a more powerful adversary. The operating principle of bandwagoning reflects the well-known idiom: "If you can't beat them, join them." The aim of appeasement is to modify the behavior of the aggressor, by offering some concession.[14]

Independent Variable: Patterns of Dependence

As noted above, dependence is defined in terms of opportunity cost incurred by actors as a result of breaking off or altering the relationship. Only if a given actor has no cost-effective means to efficiently and rapidly adjust, then the actor is considered dependent on another actor. Following this definition, the next task is to determine the conditions under which relations of dependence come to exist. No relationship between states can be accurately determined as dependence or not, however, unless at a minimum the scope and domain are first specified. As Baldwin puts it: "When one hears that a nation-state is highly dependent, the proper question is: Dependent on what actors with respect to what matters?"[15]

build up its military capability in order to ensure the containment of the aggressor state. See Mearsheimer, *The Tragedy of Great Power Politics*, 158–159.

12 Ibid., 138.

13 For more information on oil chokepoint, see EA, *World Oil Transit Chokepoints*. Available at http://www.eia.doe.gov/cabs/World_Oil_Transit_Chokepoints/Background.html.

14 For a discussion on strategies for survival see Mearsheimer, *The Tragedy of Great Power Politics*, 138–167; Organski, *World Politics*, 111–115; and Morgenthau, *Politics among Nations*, 178–197.

15 Baldwin, "Interdependence and Power: A Conceptual Analysis," 497. "No single change in scholarly writing habits would bring a more dramatic improvement in the clarity and

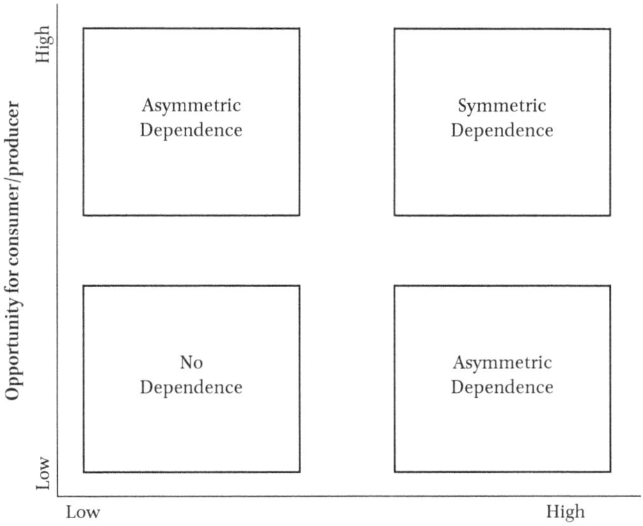

FIGURE 2.2 *Patterns of dependence*

To begin, then, it is first necessary to identify possible actors and the opportunity cost for each actor. By definition, oil rentier states are dependent on oil exports for their national income. The actors connected to the rentier state through oil relationships are consuming countries and other rentier states.[16] The magnitude of the opportunity cost determines the type of dependence between the rentier state and other actors. As shown in Figure 2.2, there are four possible patterns of dependence between the rentier state and oil consumers as well as producers.

In the case of no dependence, neither party suffers from breaking off the relationship. Therefore, the opportunity cost is low for both the rentier state and the other state (consumer or producer). Conversely, in the case of symmetric dependence all parties stand to suffer from breaking off the relationship. Therefore, neither party is in a position to exploit the relationship of dependence for its advantage. In the third case of asymmetric dependence, a situation arises in which either the rentier state is dependent on its trading

precision of such discussion than the practice of specifying who is dependent on whom with respect to what."

16 I do not consider non-rentier oil-producing states, since in these countries the government does not have a majority share in oil companies, or does not directly derive a major part of its revenues from the oil sector. Therefore, throughout the book I use the term "oil-producing countries" to refer only to oil rentier states.

partners (consumers or producers), or vice-versa. Having identified the different patterns of dependence, the next step is to determine when the opportunity cost is high for the rentier state, but not for the consumers and/or producers; and by the same token, when the opportunity cost is high for consumers and/or producers, but low for a given rentier state.

Starting with the relationship between a given rentier state and oil consumers, asymmetric dependence favors the rentier state when an oil-consuming country has no a cost-effective alternative to oil imports, while the rentier state can easily replace the consumer, or is not economically or politically injured by the loss of revenues from that consumer. For asymmetric dependence to favor oil-consuming countries, it implies that the rentier state is unable to compensate for the loss of its market share, while the consuming country has an alternative source of oil supplies. With respect to the relationship with other oil rentier states, oil price is the key issue for calculating the opportunity cost. The market price of oil affects rentier states income. Not all rentier states, however, are equally affected by the oil price. Depending on the size of their oil resources, some rentier states adjust to lower prices easier than other rentier states. Thus, the opportunity cost between two rentier states are defined based on whether or not states are able to offset low prices.

For a given rentier state to be asymmetrically dependent on another rentier state, this means it has no alternatives to compensate for loss in government income caused by lower prices. The last case is symmetric dependence or interdependence. In this case it does not matter whether the relation of dependence of the rentier state is with other producers or consumers, because all parties stand to suffer from breaking or altering the relationship.

So far I have defined the domains and patterns of dependence – i.e., over what actors the rentier state may have influence – and at the same time, what actors may have influence over the rentier state. What remains to be determined is how to assess the opportunity cost of the actors, or the conditions under which a given actor (consumer or producer) acquires influence over another actor – in other words, to assess whether a given country has the means to quickly and smoothly adjust to changes in volumes or prices without too much cost. Such an assessment would require taking into account many economic and political domestic variables,[17] as well as structural variables, such as market conditions of supply and demand and currency evaluation.

---

17 For instance, Ikenberry points out the importance of state structure in effecting different responses of states to energy crisis. See G. John Ikenberry, "The Irony of State Strength: Comparative Responses to the Oil Shocks in the 1970s," *International Organization* 40 (Winder, 1986): 105–137.

Thus, the attempt to accurately measure the opportunity cost by incorporating all relevant variables is rife with difficulties and might not be feasible. Thus, the next best thing is to single out variables which are crucial in the relationship that determines the opportunity cost.

Measuring Dependence

To assess the opportunity cost, I consider separately the relationship of dependence between a rentier state and a consumer, on the one hand, and a rentier state and a producer, on the other hand. In each case I examine the opportunity cost of the rentier state and the other actor (consumer or producer). In the case of consuming states (or a particular oil-consuming state), it is necessary to take into account the oil market conditions of supply and demand, consuming country's oil imports as percentage of total oil consumption from a given rentier state, and the relative importance of a given rentier state in the oil market. In the case of producing states the relevant indicators are the market conditions of supply and demand and the relative importance of the rentier states in the oil market.

Knowing whether there is an excess demand or supply in the market tells us a great deal about the opportunity cost involved in the relationship between rentier states and consumers. If supply exceeds demand, in most cases, asymmetric dependence favors the consumers. However, if demand exceeds supply, then it is reasonable to assume that there is no surplus in the international oil market. In that case, oil-consuming states do not have viable alternative sources for oil imports, and therefore dependence is likely to favor producers.[18] In order to simplify these measures, I use eras of boom and bust as rough indicators of the pattern of dependence between rentier states and consumers. During an oil boom period world demand exceeds world supply, therefore the boom era is an indicator of consumers' asymmetric dependence on a given rentier state. Conversely, during an oil bust period supply exceeds demand, and therefore the bust is an indicator of a given rentier state's dependence on consumers. Unlike the eras of boom and bust, in the pre-boom period relations between many rentier states and the international oil market were mediated through international oil companies. With decision-making power over supplies and prices in their hands, oil companies sought to meet growing demand at moderate prices. Thus, rentier states were dependent on oil companies (and by extension consumers). Four oil periods are shown in Figure 2.3 and discussed in detail in Chapters 3–6.

18 To be sure, however, one can examine the specific indicators of the relationship (percentage of imported oil to total consumption, and percentage of rentier state's oil exports to world exports).

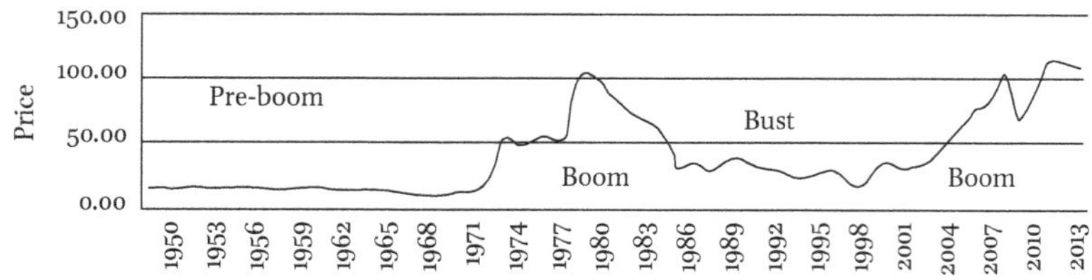

FIGURE 2.3 *Oil periods (Pre-boom, boom, bust, and boom)*
SOURCE: BRITISH PETROLEUM, STATISTICAL REVIEW OF WORLD ENERGY (ACCESSED AUGUST 2014). CRUDE PRICES IN CONSTANT DOLLARS 2013.

Although oil price is not an ideal indicator for identifying balance/imbalance in the oil market, it does suggest something about the conditions of demand and supply. In principle, market forces of supply and demand are supposed to determine price of oil, but in reality a multitude of other factors also shapes price behavior, such as non-competitive market,[19] market speculation, currency evaluation, refinery capacity, global surplus, natural disasters, and political turmoil. However, oil price is a convenient measure of the pre-boom, boom, and bust periods. As shown in Table 2.1, the boom started around 1975, when prices soared, and lasted until 1985, when prices plunged. The bust lasted until the end of the twentieth century; then a second boom period began in 2001.

The second measure is the amount of oil imported from a given rentier state as percentage of total oil consumption. There are no infallible criteria here, but countries that import 25 percent or more from a given state are vulnerable in times of a tight market. The third measure is the relative importance of a rentier state in the oil market, which is measured by the amount of oil reserves and exports. However, because oil reserves are not always recoverable it is useful to briefly describe the different types of reserves. The differentiation among various levels of resources and reserves is clarified by Ion, suggesting a pyramid-like resource classification based on ecological and economic information, moving from the general to the specific. First is the total amount of resources, which is no more than an indication of resource availability; second is reserves, defined as the portion of total resources economically and technically feasible to recover; third is possible reserves, which are the amount of resources estimated with limited geological and recovery cost; fourth is probable reserves, defined as the proportion of resources likely to be recovered with extra geological and economic information; fifth is proven reserves, which are the resources that can be recovered with certainty under current economic and

19 In actuality OPEC countries were instrumental during the 1970s and part of the 1980s in setting crude oil price, but, of course, they could not ignore levels of demand and supply.

TABLE 2.1 *Crude oil prices in US dollars, 1950–2012*

Year	Current	Constant 2013
1950	2	17
1955	2	17
1960	2	15
1965	2	13
1970	2	11
1975	12	50
1980	37	104
1985	28	60
1990	24	42
1995	17	26
2000	28	39
2005	55	65
2010	79	85
2011	111	115
2012	112	113

SOURCE: BRITISH PETROLEUM, STATISTICAL REVIEW OF WORLD ENERGY JUNE 2008.

technical information.[20] Therefore, proven reserves are only a portion of oil resources, not total resources or reserves. In sum, the adopted definition of proven oil reserves is as follows: "Proved or proven reserves are those resources known, with a high degree of certainty, to be present and that can be produced at current prices and with available technology."[21]

20 Ion, D.C., *Availability of World Energy Sources* (London: Graham and Trotman, 1980), 2–3.

21 Frank W. Millerd, "Global Oil Reserves," in Siamack Shojai, ed., *The New Global Oil Market: Understanding Energy Issues in the World Economy* (Westport, CT: Praeger, 1995), 3. A more comprehensive definition is used by the United States Department of Energy: "Reservoirs [of oil] are considered proved if economic producibility is supported by actual production or conclusive formation test, or if economic producibility is supported by core analysis, electric, or log interpretations. The area of a reservoir considered to be proved includes (1) that portion delineated by drilling and defined by gas-oil and/or oil–water contacts, if any: and (2) the immediately adjoining portions not yet drilled, but which can be reasonably judged as economically productive on the basis of available geological and engineering data. Reserves of crude oil which can be produced economically through application of improved recovery techniques, such as fluid injection, are included in this "proved" classification under certain circumstances. It is not necessary that production,

Hence, since the quality and production cost of reserves varies drastically among oil-producing countries, not all reserves are sufficiently developed for exports. For instance, Canada's reserves are second only to that of Saudi Arabia, but the quality and cost of developing reserves are a major constraint on production capacity. In other words, Canada's production capacity is not proportional to its reserves. Finally, because of high domestic consumption, despite large reserves and production output in many producing countries, the amount of oil exported is low. Therefore, the proven reserves indicator alone is a necessary but not sufficient criterion. Also needed is export capacity, which is a straightforward indicator of the amount of oil exported to the international market.

In the case of dependence between a given rentier state and other rentier states, the market condition of high supply and low demand renders some rentier states vulnerable to other oil-producing states. In a glut market rentier states earn less income due to low oil prices. However, rentier states with large reserves and export capability are likely to suffer less by producing more oil. On the other hand, small rentier states lack the surplus capability to offset low prices. Based on reserves and export capabilities it is possible to conclude the following about the opportunity cost between producers: changes in oil prices generate less opportunity cost for the producer with larger reserves and greater export capacity.

Causal Mechanisms

Revenue Mechanisms

When dependence vis-à-vis the consumers shifts in favor of a given rentier state, such a state acquires new resources to invest in strategies of security. Higher demand for oil and higher prices provide rentier states with more income. With more income the rentier state affords to increase spending on defense and internal security apparatuses (internal balancing), on foreign aid (external balancing), and on social welfare services and goods (internal validation). By the same token, during the glut, soft demand and low price translate into less income for the rentier state. Less wealth implies that the rentier state is forced to cut spending on internal validation and internal balancing.

Supply Mechanism

Under the condition of a tight market the gap between world production and world consumption is reduced, decreasing excess capacity. Consequently, a little excess capacity implies that consuming states become vulnerable to oil

gathering, or transportation facilities be installed or operative for a reservoir to be considered proved." Quoted in Millerd, "Global Oil Reserves," 3.

disruptions and higher prices. Therefore, the tight market endows the rentier state with an opportunity to exploit oil resources through strategies of external validation and external balancing. Both strategies are shaped by oil resources and can take different forms. In the case of external validation, for instance, the rentier state can initiate an embargo or cut back exports in order to increase its legitimacy and mollify domestic discontent at home. As for external balancing, the rentier state can balance with consuming countries by maintaining sufficient oil supplies in the market. As Dawisha points out:

> It was again Saudi Arabia who stood almost virtually alone against moves by the other members to institute a co-ordinated programme of cutbacks in oil production in order to maintain the price, and to index-link oil prices so as to maintain their value in real terms.[22]

Price Mechanism

When market conditions shift from scarcity to glut, relations of dependence between rentier states and consuming states shift in the favor of the latter. The excess supply means the rentier state is in a less-likely position to employ external validation and external balancing. Moreover, the soft market impacts relations with producers, which in turn bring about changes in the rentier state's capabilities for confronting internal and external threats. Because oil prices are influenced by demand as much as by supply, it matters for a given rentier state how much oil is produced by other rentier states – given that a higher output leads to further decline in oil prices. Thus, states with higher production capacity are more likely to gain influence over states with less production capacity. States with high production capacity, in other words, acquire the ability to employ strategies of external validation and external balancing. To bring about a change in the behavior of other rentier states, a given rentier state with relatively large reserves has the advantage of producing more oil, causing the price to plunge. Although low prices are harmful to all oil rentier states, in the short term, states with larger reserves suffer less. Their higher output compensates for losses caused by lower prices. The advantage of large reserves and production capacity is most effective under the condition of a weak market, when prices are already low.

Conclusion

In this chapter I presented a theoretical framework termed neo-rentier theory. Neo-rentier theory seeks to explain Saudi Arabia's domestic and international

22 Dawisha, "Saudi Arabia's Search for Security," 28.

security strategies. Neo-rentier theory is based on a modified logic of rentier theory. It retains the definition of the rentier state but refines and increases its explanatory power by incorporating assumptions and insights from realism and liberalism. From realism it borrows the assumption of survival as the most important goal. The survival goal, however, is broadened beyond the realist conception of external military threat. Two other types of threats are also taken into consideration: political threats as well as internal threats.

By the same token, neo-rentier theory departs from realism on the conception of power. Rather than conceiving power only in terms of military capabilities, the theory incorporates the liberal view of dependence as a power resource. Although the liberal tradition has viewed economic dependence as an impetus of cooperation and peace, a number of scholars have convincingly made the case for dependence as a source of power. As a source of power, the relation of oil dependence is analyzed in terms of the opportunity cost of severing or altering the relationship. In order for actor A to gain influence over actor B, the opportunity cost of severing the relationship for A has to be insignificant, but for B has to be significant. In that case the relationship between A and B is defined as asymmetric dependence.

In these terms, dependence is used in neo-rentier theory as an independent variable. Accordingly, the variation on the independent variables indicates whether the rentier state is dependent on the other states (consumers or producers) or the consumers or other producers are dependent on a given rentier state. The dependent variable is the variation in responses to internal and external threats. Responses to threat, however, do not only vary in form but also in the level of intensity. This is a crucial point, which is often overlooked in rentier theory literature. The general hypothesis is that the pattern of dependence conditions the types and intensity of security strategies available for a rentier state to choose from in response to internal and external threats.

CHAPTER 3

The Pre-Boom (1950–1970)

The transformation from a pre-oil economy to an oil-rentier economy improved the economic and military capabilities of the Saudi state. In the pre-boom era, however, oil market conditions and the oil concession constrained the role of oil dependence as a power resource. This chapter examines oil dependence and Saudi security behavior during the pre-boom. The chapter is divided into three sections. Section one deals with the two measures of dependence, market forces of supply and demand, and Saudi Arabia's levels of export and proven reserves. While in section two strategies of internal validation and internal balancing are evaluated, section three examines security strategies employed in response to major security concerns.

Pattern and Domain of Dependence in the Pre-Boom Era, 1950–1970

Oil Market: Supply and Demand

The decades that followed the Second World War were marked by a rapid economic growth. Between 1953 and 1975 the growth rate of world manufacturing industries averaged 6 percent.[1] In large part this dramatic growth was caused by several factors:

> the recovery of war-damaged economies, the development of new technologies, the continued drift from agriculture to industry, the harnessing of national resources within "planned economies," and the spread of industrialization to the Third World.[2]

To support this rapid growth in manufacturing, demand for energy supplies drastically increased. The rapid shift from coal to oil, together with growth in automobile ownership and expansion of air travel, stimulated a surge in demand for oil. Between 1960 and 1970 the world consumption grew at a remarkable average of 12 percent a year (see Table 3.1).

1 Paul Kennedy, *The Rise and Fall of the Great Powers: Economic Change and Military Conflict from 1500 to 2000* (New York: Random House, 1987), 414.

2 Ibid.

 | DOI 10.1163/9789004277731_005

TABLE 3.1 *World oil consumption (Million b/d), 1960–1970*

Year	Total OECD	Total	World
1960	16	6	21
1961	17	6	23
1962	18	7	25
1963	20	7	27
1964	21	8	29
1965	23	8	31
1966	25	9	34
1967	26	10	36
1968	29	10	39
1969	32	11	43
1970	35	12	47

SOURCE: ENERGY INFORMATION ADMINISTRATION (EIA), ANNUAL ENERGY REVIEW 2007.

On the supply end, oil production also increased in many countries. In Table 3.2 shows the expansion in production by regions in the period 1960–1970. The world production grew from 1960 to 1970 by 125 percent, marking an average annual growth rate of 12.5 percent. Although the world production steadily increased, most of the increment came from non-Western countries in Africa, Asia and the Pacific, as well as OPEC countries. Between 1960 and 1970 production rose in Africa by 2037 percent, Asia and the Pacific by 180 percent, Saudi Arabia by 189 percent, and OPEC countries (excluding Saudi Arabia) by 125 percent. For that same period, the growth rate of world production averaged 12.50 percent a year. As a result of over-production, prices became very cheap; this in turn encouraged consumption. In constant 2007 dollars prices fell from $15 per barrel in 1950 to $10 per barrel in 1970.

Saudi Arabia: Oil Reserves and Exports

By 1960 Saudi Arabia had become one of the most important oil exporters in the world. By the end of the 1960s its oil exports had tripled (see Table 3.3). In 1970 exports from Saudi Arabia amounted to more than exports of North America, Eastern Europe, Western Europe, and Asia and the Pacific. Between 1955 and 1970, Europe, on the one hand, and Asia and Australia, on the other hand, were the two most important destinations of Saudi oil exports. Out of

TABLE 3.2 *Oil production by region (1000 b/d), 1960–1970*

Region/Country	1960	1962	1964	1966	1968	1970
North America	7,599	8,060	8,431	9,252	10,205	10,963
Latin America	3,725	4,246	4,494	4,573	5,023	5,175
Eastern Europe	3,207	3,971	4,725	5,557	6,426	7,235
Western Europe	306	354	412	434	474	470
Saudi Arabia	1,314	1,643	1,897	2,602	3,043	3,799
Africa	282	808	1,704	2,813	3,962	6,032
Asia and Pacific	551	635	696	870	986	1,544
Middle East[1]	4,017	4,613	5,785	6,787	8,233	9,980

(1) Saudi Arabia is not included.
SOURCE: OPEC, ANNUAL STATISTICAL BULLETIN, 2007.

TABLE 3.3 *Oil exports by region (1000 b/d), 1960–1970*

Region/Country	1960	1962	1964	1966	1968	1970
North America	124	255	280	343	462	676
Latin America	2,140	2,375	2,504	2,409	2,597	2,572
Eastern Europe	358	524	725	1,006	1,169	1,323
Western Europe	21	41	17	13	16	42
Saudi Arabia	1,081	1,379	1,604	2,277	2,643	3,217
Africa	222	753	1,587	2,630	3,775	5,852
Asia and Pacific	366	393	373	400	587	859
Middle East[1]	3,243	3,808	4,866	6,119	7,176	8,926

(1) Saudi Arabia is not included.
SOURCE: OPEC, ANNUAL STATISTICAL BULLETIN, 2007.

total Saudi oil exports 33 percent to 44 percent was destined to Europe, and 30 percent to 40 percent was exported to Asia and Australia.[3]

The dramatic increase in Saudi exports was paralleled by similar growth in the volume of proven oil reserves. By 1960 the kingdom held the largest amount of proven reserves in the world (see Table 3.4).[4] While in some other regions proven

3 SAMA, *Annual Statistical Bulletin*, Various Years. Accessed 2008. http://www.sama.gov.sa/sites/SAMAEN/Pages/Home.aspx.

4 Proven reserves are defined as those resources that are commercially and technically viable to recover.

TABLE 3.4 *Proven oil reserves by region as a percentage of world reserves, 1960–1970*

Region/Country	1960	1965	1970
North America	12.6	11	9.1
Latin America	8.6	7.1	4.8
Eastern Europe	8.5	10.2	11.1
Western Europe	0.6	0.7	1.3
Saudi Arabia	18.2	18.7	25.8
Africa	2.7	6.6	9.3
Asia and Pacific	4	3.7	3.1
Middle East[1]	44.7	42	35.5

(1) Saudi Arabia is not included.
SOURCE: OPEC, ANNUAL STATISTICAL BULLETIN, 2007.

reserves fell, Saudi Arabia's share of world reserves increased from 18.2 percent (1960) to 25.8 percent (1970). The rapid development and massive expansion was largely made possible because of the relative low cost of capital investment required in exploration and development of Saudi oil reserves. As Parra points out, during the period from 1952 to 1973 the United States' share of the world total capital expenditures on oil and gas production amounted to $91 billion out of a world total of $137 billion, but made an insignificant addition to the world proven reserves. In comparison, during the same period, the Middle East contributed 42 percent of the increase in world production at a cost of only $6 billion.[5]

Patterns of Dependence during 1950–1960

Despite its large reserves and production capacity, Saudi Arabia did not enjoy asymmetric dependence vis-à-vis the consumers or the producers. The reason is partly the availability of abundant supplies. In terms of supply, the production capacity expanded in other producing countries in order to keep up with growing demands. New additions of oil supplies were brought to market from countries like Venezuela, Mexico, Canada, and the United States. The second reason is the limitation of the concession, which delegated all decision-making power over oil prices, production, and exports to ARAMCO. Thus, what mattered for consumers and other producers were the decisions taken by ARAMCO and other oil companies, not the Saudi state. At best the Saudi state had some leverage on production cutback, but did not have the capacity to overproduce

5 See Francisco Parra, *Oil Politics: A Modern History of Petroleum* (London: I.B. Tauris & Co. Ltd, 2004), 40.

or pressure prices. At the same time, ARAMCO contributed in the form of royalty and profit sharing to the Saudi state national income. The increasing reliance on oil export revenues for economic development and political stability rendered the Saudi state dependent on the consuming states. On the other hand, without leverage over prices or supplies, coupled with availability of supplies in the market – the consuming states were not dependent on the Saudi state. In other words, the opportunity cost was low for the consuming states but high for the Saudi state. Under these circumstances, Saudi Arabia could not easily use internal balancing and validation as well as external validation and balancing.

The state income considerably improved as an outcome of multiple changes in the oil concession and higher oil production. First, major developments in the oil sector expanded the Saudi oil production, which in turn corresponded with higher return for the Saudi state and ARAMCO. The construction of the Trans-Arabian Pipeline (Tapline), which begun in 1947, was finished in 1950. The pipeline, stretching over a distance of 1,068 miles from the eastern region of Saudi Arabia to Sidon in Lebanon, had an initial capacity of 300,000 barrels per day, but increased in the year 1957 to 2450,000 barrels per day. A new refinery was built at Ras Tanura to handle 50,000 barrels a day, increasing to 200,000 during the 1960s. At the same time, in response to surging global demand for oil in the 1950s and 1960s, ARAMCO increased the rate of production.

Second, the Saudi government was able to boost its revenue by renegotiating its share of profit on several occasions. In 1951 the government introduced the 50–50 profit sharing agreement, which drastically increased the level of royalty from 22 cents a barrel in 1950 to 1980–90 cents a barrel in the mid-1950s. Again, in 1962, the government pressed for a higher allowance from ARAMCO and after two years of difficult negotiation an agreement was reached.[6] The ARMACO offer was a three-year package in which government royalties increased gradually, amounting to an extra 4.5 cents a barrel to 5.0 cents a barrel.[7] Table 3.5 shows the substantial increase in oil revenue and the centrality of oil royalties in government revenue.

Third, the Saudi government signed an agreement with ARAMCO in 1966, which stipulated that all oil revenues were to be calculated on the basis of posted price, not realized price.[8] The realized price compelled the Saudi

6 It was the first bargaining that took place between OPEC and the oil companies. OPEC was founded with five member states: Venezuela, Saudi Arabia, Iran, Iraq, and Kuwait. For political and economic reasons, however, the role of OPEC throughout the 1960s was very limited. ARMACO continued to exercise power over pricing, production, and exports.

7 Seymour, *OPEC Instrument of Change*, 49.

8 The realized price is the price which ARAMCO charged its affiliates and customers.

TABLE 3.5 *Saudi Arabia revenue, 1954–1970*

Fiscal year	Total revenue (Million SR)	Oil revenue (Million SR)	Percentage of total revenue
1954–55	1,355	966	71
1957–58	1,498	1,241	83
1958–59	1,410	1,145	81
1959–60	1,405	1,149	82
1960–61	1,786	1,410	79
1961–62	2,166	1,682	78
1962–63	2,452	1,951	80
1963–64	2,686	2,284	85
1964–65	3,112	2,607	84
1965–66	3,961	3,179	80
1966–67	5,025	3,987	79
1967–68	4,937	3,559	72
1968–69	5,536	4,243	77
1969–70	5,966	5,261	88

SOURCE: MINISTRY OF ECONOMY AND PLANNING, ANNUAL STATISTICAL BOOK, 1965 AND 1970.

government to shoulder with ARAMCO the burden of market-adverse effects. Under the new agreement, ARMACO continued to be in charge of oil pricing, but with the establishment of the posted price the government was sheltered from financial instability caused by the realized price. In other words, the oil companies had to absorb the cost and the benefit of market movements. In light of these arrangements, the financial position of the kingdom improved, making it possible for the state to allocate a limited amount of funding to survival strategies.

Internal Validation and Internal Balancing in the Pre-Boom Era

Internal Validation

Despite the steady flow of oil revenues Ibn Saud (1902–1953) spared very little oil revenue on societal development, apart from the subsidies and grants extended to tribal chiefs. His successor, the eldest son Saud, made no significant budget changes. King Saud funneled much of the oil revenues to his personal usage, building lavish places and leading an extravagant lifestyle.[9] It was not until his half-brother Crown Prince Faisal took control that a policy of internal validation emerged.[10] Helped by oil resources, Faisal adopted a course of actions aimed at reinforcing the legitimacy of the royal family at home. In 1962 a modernization program was initiated with a decree of the ten-point program. Points six and nine were concerned with duties of the government toward its citizens, especially the improvement of living standards.[11] These two points enumerated various government programs and initiatives, such as free health and education, subsidies to industries, and additional allocations to development projects.[12] Although the range and the magnitude of such services were still at a nascent stage, they were far more than what the government had offered in previous decades, which was little more than order and security. The intended purpose of the provision of socio-economic services was to increase the base of legitimacy. In effect the state was forging a secondary accord with the society in which political loyalty and acquiescence were exchanged for economic and social rewards.

New institutions were set up to oversee the development of social services, including health, education, transportation, and religious affairs. For instance, nine new ministries were established with specialized directives and budget allocation. These are the Ministry of Education (1953), Ministry of Health (1953), Ministry of Agriculture (1953), Ministry of Transportation (1953), Ministry of Commerce and Industry (1954), Ministry of Petroleum (1961), Ministry of Pilgrimage (1962), and the Ministry of Information (1963). The budget allocation to these ministries (including salaries, general expenditures, other expenditures, and projects) jumped from 658 million SR in the fiscal year 1965–1966 to 931 million SR in the fiscal years 1969–1970. The creation of a highly centralized bureaucracy, with diversified and specialized ministries, agencies, and departments greatly enhanced the ability of the state to penetrate and control the various activities of the society.

---

9 By 1958 the Saudi government was in a financial crisis, with a debt amounting to $480. Largely the crisis was a result of Saud and his government's extravagant spending and unsound financial regulations. See Holden and Johns, *The House of Saud*, 199.

10 During much of the 1950s and early 1960s, King Saud and his half-brother Crown Prince Faisal were locked in a power struggle. Because of his incompetent leadership, Saud was forced to abdicate under strong pressure from senior princes and the *Ulama*. Full transfer of power from Saud to Faisal took place in 1964.

11 Interestingly enough there was no mention of obligations toward the state.

12 For a full text of the ten-point reform program, see Gerald de Gaury, *Faisal: King of Saudi Arabia*, (London: Arthur Barker Limited, 1966), 147–151.

The civilian bureaucracy in itself served as a vehicle for employment. For instance, in the fiscal year 1965–1966, the nine ministries collectively added approximately 66,460 jobs. The number is even higher after including jobs from the Ministry of Interior, which nearly doubled, from 7245 in 1959 to 12790 in 1965. In contrast to 1950, with only a few hundred civil servants, by 1965 the civilian bureaucracy had 101909 positions. Expenditures on salaries of all civil servants, as shown in Table 3.6, represented more or less one-third of total revenues and even higher as percentage of oil revenues.

The government's role in the economy also drastically grew in more than one way. It enacted new market regulations, increased spending on development projects, and established initiatives for expansion of the public sector. For instance, as shown in Table 3.7, as a percentage of oil revenues, allocation to development projects was tripled by 1968–1969. In order to increase its grip on the private sector the government employed chambers of commerce for assistance on a variety of economic issues, including advising the government, gathering information on imports, and the implementation and enforcement of policies. In return for their services, the government provided them with significant subsidies.[13]

TABLE 3.6 *State expenditures on salaries and compensations, 1959–1969*

Fiscal year	Total (Million SR)	Percentage of total revenue	Percentage of oil revenue
1959–60	411	29.25	35.77
1960–61	583	32.64	41.35
1961–62	577	26.64	34.3
1962–63	651	26.55	33.37
1963–64	1,017	37.86	44.53
1964–65	1,112	35.73	42.65
1965–66	1,255	31.68	39.48
1966–67	1,370	27.26	34.36
1967–68	1,433	29.03	40.26
1968–69	1,600	28.9	37.71

SOURCE: COMPILED BY THE AUTHOR FROM MINISTRY OF ECONOMY AND PLANNING, ANNUAL STATISTICAL BOOK, 1965 AND 1970.

13 According to Chaudhry, chambers of commerce began receiving subsidies in 1961, starting with the Riyadh Chamber of Commerce. In 1969 the Chambers of Jeddah, Riyadh, and Dammam were receiving an annual subsidy of SR 750,000. For more information see Chaudhry, *The Price of Wealth: Economies and Institutions in the Middle East*, 88.

TABLE 3.7 *State expenditures on development projects, 1960–1969*

Fiscal year	Total (Million SR)	Percentage of total revenue	Percentage of oil revenue
1960–61	2,05,542	11.5	14.6
1961–62	2,89,242	13.4	17.2
1962–63	3,77,091	15.4	19.3
1963–64	4,43,651	16.5	19.4
1964–65	7,74,191	24.9	29.7
1965–66	9,33,644	23.6	29.4
1966–67	11,45,120	22.8	28.7
1967–68	17,47,208	35.4	49.1
1968–69	19,64,701	35.5	46.3

Projects included: education, health, communication and transportation, agriculture and water, and unspecified projects.
SOURCE: COMPILED BY THE AUTHOR FROM MINISTRY OF ECONOMY AND PLANNING, ANNUAL STATISTICAL BOOK, 1965 AND 1970.

The private sector was driven toward dependence on the government. Massive spending on development projects served as another channel of oil wealth redistribution in the form of rewarding lucrative contracts to the private sector. State-owned corporations were set up and allocated subsidies, playing different but active roles in the economy. For instance, the General Petroleum and Minerals Organization (Petromin) was mandated with the development of commerce and industrialization in the area of petroleum; the Saudi Airline was overhauled and expanded to cover many destinations around the world; the Railroad organization was to oversee the development of a large-scale road-building program.[14]

Internal Balancing

The capability of the state improved, because the oil concession between the Saudi state and ARAMCO was beginning to pay off politically and economically. Whereas ARMACO, backed by the United States government, enjoyed full control over Saudi oil, the territorial integrity and the Saudi state, as well as the reign of the Saudi royal family, became part and parcel of the United States oil interest in the region. Acting on its own interest, as much as on the interest of

14 Total government subsidies in the period of 1965 to 1969 were: 41 million SR for Railroad Organization; 10 million SR for Petromin; 210 million SR for Saudi Airline.

the Saudi royal family, the United States reversed its policy of ambivalence and became directly involved in the security of the kingdom.[15]

In 1950 the United States responded to the plea of King Ibn Saud for military assistance by abandoning the year-to-year arrangements and endorsing a more substantial agreement.[16] Although short of a full-fledged military alliance with the United States, King Ibn Saud was offered and accepted a security package, which provided military and financial assistance.[17] In 1951 the two countries signed the Mutual Defense Assistance Agreement (MDAA), which laid down the foundation for a close relationship based on arms sales and military training that has grown over the years into a price tag worth many billion dollars.[18] As part of the MDAA, King Ibn Saud agreed to extend the lease on the Dhahran airfield for five years, while the United States began to supply military equipment and deployed a military training mission to build up the Saudi armed forces.

The American training mission was formally reestablished in 1953 as the American Military Training Mission (USMTIM), which has become the cornerstone of the United States-Saudi Arabia military relationship. The USMTM was charged with all aspects of building up modern Saudi armed forces, as stated in the USMTIM Mission Statement:

> USMTM was chartered to assist and advise the Saudi Arabian Armed Forces with respect to the building of military equipment, plans,

15 Consider for instance, the statement by Assistant Secretary of State George McGhee that the United States "will take most immediate action at any time that the integrity and independence of Saudi Arabia is threatened." Quoted in Safran, *Saudi Arabia: The Ceaseless Quest for Security*, 66.

16 According to Safran the offer consisted of five components. The Saudi part of the deal included: first, treaty of friendship, commerce, and friendship; second, support from American technicians; third, loans from the American Export–import Bank; fourth, military aid program including arms and training. For its part the United States was granted a long-term lease on the Dhahran airfield; see Safran, *Saudi Arabia: The Ceaseless Quest for Security*, 66.

17 The offer was accepted with one caveat – the Dhahran lease was to be extended for five years rather than a longer-term arrangement.

18 According to a study conducted by the Congressional Research Service, from 1950 through 2006 Saudi Arabia purchased from the United States military weapons, military equipment, and related services worth more than $62.7 billion, approximately 19 percent of all foreign military sales. Additionally, during the same period, the Kingdom purchased foreign military construction services (FMCS) worth over $17.1 billion, amounting to 85% of all FMCS sales. See Christopher M. Blanchard, Saudi Arabia: Background and U.S. Relations, CRS Report for Congress. Accessed April 2008. http://www.fas.org.

> organization, administrative procedures, training methods, and the conduct of such training.[19]

As a result of the USMTIM mandate to modernize the Saudi armed forces, the kingdom acquired modern military equipment including aircrafts, armored cars, tanks, and the pricey $300-million air defense system.[20]

Increment in oil revenues led to greater expenditures on defense and security. The accrued oil revenue relieved the state from the cash deficit burden that is typical of Third World countries, but also provided the Saudi state substantial financial resources to shape its security environment. It facilitated spending on the creation and upgrading of external defense capability and internal security apparatuses. Figures in Table 3.8 show that total spending steadily increased in the 1960s, with a considerable jump in the fiscal year 1965–1966. The bulk of the security budget allocation, however, went to the Ministry of the Interior. For fiscal years 1965–1970, in comparison to the 12–20 percentage allocated to the Ministry of Defense, 70–80 percent was allocated to the Ministry of the Interior, and even a little higher if other security apparatuses were included, such as the Intelligence Bureau, the Ministry of Information, and National Guards. The figures also provide a good insight into the security concern of the kingdom and the manner of responses adopted by the royal family. The government invested heavily in the internal security apparatuses, mainly because the Saudi state was far more concerned about internal instability. External security for the most part was outsourced to the United States.

Cases of Major Threats

While the previous section examined the intensity of internal validation and internal balancing, this section considers major threats to the kingdom in the period 1950–1970. It examines the type of strategies that were adopted by the Saudi state in response to the Suez Crisis, Nasser's Pan-Arabism, the Yemen Civil War and Nasser's military intervention, and the 1967 Six-Day War.

19 The online version of the Mission Statement is available from The United States Training Mission (USMTM) to the Kingdom of Saudi Arabia. Accessed April 2008. http://usmtm.org/index.html.

20 For more information on the type of military equipment Saudi Arabia purchased during the 1950s and 1960s, see Safran, *Saudi Arabia: The Ceaseless Quest for Security*, 103–112.

TABLE 3.8 *Defense and internal security allocations (Million SR), 1950–1970*

Security Expenses	1950–1951	1954–1955	1957–1958	1958–1959	1960–1961	1962–1963	1965–1966	1967–1968	1968–1969	1969–1970
Ministry of Information	–	–	–	–	–	–	75	107	100	99
National Guards	–	–	60	55	55	61	188	202	234	234
Royal Guards			16	11	14	16	18	18	22	–
Intelligence Bureau	–	–	–	–	–	–	15	16	16	18
Ministry of Interior	12	82	87	107	121	183	3,473	3,859	4,561	5,651
Ministry of Defence	100	472	310	196	180	278	541	1,248	1,217	1,587
Total	112	554	473	369	370	538	4,310	5,450	6,150	7,589

SOURCE: COMPILED BY THE AUTHOR FROM MINISTRY OF ECONOMY AND PLANNING, ANNUAL STATISTICAL BOOK, 1965 AND 1970.

Case One: The Suez Crisis (1956)

In 1956 President Gamal Abdel Nasser nationalized the Anglo-French company that owned the Suez Canal, precipitating the Anglo-French-Israeli attack on Egypt. The Canal served as a strategic oil passage to Western Europe, through which almost two-thirds of Europe's oil supplies from the Middle East passed in 1955. The alarm triggered by the nationalization was clearly expressed in the words of British Prime Minister Sir Anthony Eden: "Nasser can deny oil to Western Europe and we here shall all be at his mercy."[21] In the Arab world, however, Nasser's decision to nationalize the Canal was widely welcomed. In Damascus over 100,000 rallied in the street in a show of solidarity with Egypt; in Jordan public opinion was fully on the side of Egypt; and in Riyadh demonstrators hailed Nasser as the vanguard of Arabism. Throughout the Arab world a general strike took place on 16 August 1956 to protest the opening of the international conference on the nationalization of the Canal. The Arab leaders, too, expressed public support for Egypt. The Council of the League of Arab states expressed support for Egypt and approved the nationalization of the Canal.[22]

The crisis finally erupted when Israel, Britain, and France together launched an attack on Egypt. The tripartite attack on Egypt provoked a widespread reaction in the Arab world. Anti-Western demonstrations broke out in Libya, Syria, Jordan, Bahrain, and Iraq. Oil fields in Kuwait were sabotaged and pumping stations on the IPC pipeline from Iraq to the Mediterranean were destroyed. In Saudi Arabia a French-run factory was attacked. Two months later, however, the crisis came to an end. After heavy pressure from the United Nations and the United States, in December 1956, Britain and France withdrew, and three months later Israel evacuated the Sinai.

Responses: External Validation

As public opinion at home and throughout the Arab world rallied against the attack on Egypt, the Saudi regime was in no position to do anything but demonstrate support for Egypt. To protect itself from internal instability and hostility from neighboring Arab countries, the regime took on the following measures: first, it broke off diplomatic relations with Britain and France; second, British and French tankers were banned from loading oil at Saudi ports; third, other nations' tankers were also banned from carrying Saudi oil to Britain and France; and fourth, oil supplies were cut off to the British protectorate of

21 Quoted in James Bamberg, *British Petroleum and Global Oil, 1950–1975: The Challenge of Nationalism* (Cambridge: Cambridge University Press, 2000), 81.

22 James P. Jankowski, *Nassers' Egypt, Arab Nationalism, and the United Arab Republic* (Boulder: Lynne Rienner Publishers, 2002), 83.

Bahrain. However, the rest of the world continued to receive Saudi oil. Therefore, it was possible for the international oil companies to divert tankers to British and French ports. But this diverted oil came at a much higher cost than before, since a transportation premium was added to the oil price. With the closure of the Canal and the damage to the IPC pipeline from Iraq to the Mediterranean, oil shipments destined to any European ports had to make a long journey around the Cape of Good Hope. The 1956 Suez Crisis marks the enactment of the first Arab oil embargo against Western countries.

Case Two: Nasser's Pan-Arabism (1958–1962)

During the 1950s and 1960s a new era of inter-Arab politics was ushered in, so intense with subversion and intrigue and propaganda warfare that it was dubbed by Kerr as "the Arab cold war." It was a cold war that pitted the "revolutionary" regimes under Nasser's leadership against the "reactionary" and "moderate" regimes under the leadership of Saudi Arabia.[23] Right after the Suez crisis, Gamal Abdel Nasser emerged as the single most critical danger to the Saudi royal family. Having snatched political victory from the jaws of military defeat during the Suez crisis in 1956, Nasser won public opinion throughout the Arab world. His prestige and popularity was unparalleled in the Arab world, and his call for pan-Arabism appealed to both the elites and masses. The combination of these two factors made the Nasser challenge extremely dangerous to the Saudi regime. To survive, the ruling family had to withstand the mounting internal and external threat posed by the Egyptian President.[24]

Nasser's pan-Arabism embodied a set of ideas, which were anathema to the ethos of the Saudi royal family. Where Arabism meant progress and change, the Saudi royal family aimed to preserve tradition and continuity; where Arabism abhorred the system of monarchy, the Saudi royal family was nothing but an absolute monarchy; where Arabism demanded breaking off diplomatic relations with Western countries, the Saudi royal family was economically and politically dependent on the West by the necessity of security and oil revenue. Thus, the infiltration of Arabism into the kingdom and its appeal to the masses, especially the educated segment of the society, posed a major security challenge.

---

23 It needs to be noted that tense rivalry took place between two revolutionary regimes at various times: Egypt and Iraq, Egypt and Syria, and Syria and Iraq. Yet these regimes were on the same ideological side against Arab monarchies, including Saudi Arabia, Jordan, Morocco, and Iran.

24 In the midst of an external crisis, an internal crisis was unfolding between King Saud and his half-brother Prince Faisal.

Moreover, the call for unity among Arab states embodied in pan-Arabism was not to be taken lightly by the royal family. The Saudis correctly read in the call for unification two kinds of threat, an internal as well as external. With Cairo denouncing the monarchy as a stumbling block on the road of Arab unity the Saudi royal family feared an internal rebellion. Eliminating the "reactionaries" and "backward" regimes was considered a revolutionary and necessary forward step toward Arab unity. Among those who heeded the call for change were members of the armed forces, and some members of the ruling family. After renouncing his title, Prince Talal announced plans "to establish a national democratic government and to leave the people free to choose the kind of government they prefer."[25] The armed forces were also infiltrated with supporters of Nasser. In the early days of the coup d'état in Yemen, several Saudi aircraft with military supplies destined for the Yemeni royalists defected to Egypt.

As for the external threat, the Saudi regime feared being encircled by hostile regimes. Since the foundation of the kingdom in 1932, Saudi Arabian foreign policy had aimed to prevent union among Arab states for this very reason. Thus, when Syria and Egypt established a union under the command of Nasser in February 1958, Saudi Arabia became extremely nervous. The union with Syria was particularly alarming, because of Syria's geographic proximity. King Saud's attempt to foil the union by financing a coup d'état was exposed by Colonel Abdel Hamid Sarraj, director of Syrian Intelligence. Sarraj revealed the details of the plot including checks worth £1.9 million made by King Saud as a payment on a total of more than £120 million to be paid after the assassination of President Nasser. The blunder of King Saud exacerbated the tension between Cairo and Riyadh.[26]

Responses: Appeasement

Initially, the threat of pan-Arabism was confronted with an accommodative policy. Upon assuming full statutory power, Crown Prince Faisal not only expressed his support for Nasser's pan-Arabism, but also indicated his willingness to bring Saudi Arabia into the United Arab Republic union. In order to defuse nationalist sentiments, Faisal sought to broaden the legitimacy of the royal family by neutralizing the opposition. Measures were taken to distance the kingdom from the United

---

25 Holden and Johns, *The House of Saud*, 227.

26 In Sadat's words, "He [King Saud] had financed the breakup of the union with Syria, and led the campaign against Egypt, while his country had common borders with the Yemen...I convinced the [Presidential] council of the necessity for supporting the Yemeni Revolution, and we did so." See Anwar Sadat, *In Search of Identity* (London: Collins, 1978), 162.

States as well as anti-Nasser Arab regimes, namely Lebanon and Jordan. Faisal reversed the policy of rapprochement with Jordan, downgrading the relationship between the two kingdoms from cooperation to neutrality. In stark contrast to Saudi Arabia's support of Jordan during the 1957 crisis when Saudi troops were put under the command of King Hussein to defend his reign against the threat of Nasserism, Faisal in 1958 withdrew support of Jordan. Saudi troops were evacuated from Jordan and the subsidy of five and a half million sterling to Jordan was cancelled. Moreover, in the wake of the Iraqi coup which left Jordan cut off from its oil needs, Saudi Arabia was unwilling to grant airlift access over its territories to Jordan, for fear of upsetting the tacit understanding with Nasser.[27]

In its effort to accommodate Nasser, Saudi Arabia also adopted a cool relationship with the United States; straining, if not reversing, the fundamental security pact with the United States dating back to the late 1940s. The setback in Saudi-American relations was manifested in the wake of the Iraqi coup. A few days after the Iraqi coup, American and British troops were dispatched to Lebanon and Jordan to buttress security of these countries against internal and external threats. Saudi Arabia was confronted with two courses of action, either lining up with Nasser or with the United States. Weighing his options, Faisal sided with the more risky threat. Not only did Faisal label the American and the British intervention in Lebanon and Jordan as "aggressive," but he also promised not to renew the lease for the Dhahran Airfield after its expiration. King Saud upheld that promise, canceling the Dhahran Airfield Agreement in May 1961.[28] Moreover, Saud proclaimed his support for Arab nationalism, and made promises to Nasser not to renew the Dhahran lease.[29] In the Council of Ministers, Saud appointed liberal princes and progressive non-royal technocrats with sympathetic views toward Nasser.[30]

---

27 The Iraqi coup brought down the Hashemite monarchy and brought to power nationalist army officers believed to be pro-Nasser. The Presidency of the Council of Ministers "denied that Saudi Arabia had given permission for US planes to fly over Saudi territory carrying oil from Bahrayn to Jordan." Chronology, *Middle East Journal* (July 1 – September 15, 1958), 443. The airlift to Jordan was resumed after Israel lifted its ban on flights over its territories. Chronology, *Middle East Journal* (July 1 – September 15, 1958), 443.

28 Chronology, *Middle East Journal* (Summer 1961), 319.

29 For instance, Saudi Deputy Minister of Foreign Affairs, Yusuf Yasin, announced, "Saudi Arabia and the UAR are allies whose policy is first to promote Arab interests." Chronology, *Middle East Journal* (Winter 1959), 93. It was also reported that Crown Prince Faisal claimed, "Saudi Arabia has achieved an independent foreign policy based on neutralism and Arab nationalism." Chronology, *Middle East Journal* (Winter 1959), 93.

30 Prince Talal was appointed as Minister of Finance and National Economy, Prince Abdul Mohsen as Minister of Interior, and Prince Badr as Minister of Communications. Abdullah Tariki, a commoner, was appointed as Minister of Petroleum and Mineral Resources.

Case Three: Yemen Civil War and Nasser's Military Intervention

The second stage started with an Egyptian intervention in North Yemen in 1962, and lasted until the disastrous defeat of Egypt in the 1967 War at the hands of Israel. In the wake of the coup which brought down Imam Muhammad al-Badr of Yemen in September 1962, a civil war broke out which divided the country between royalists aided by the Saudi ruling family against republicans aided by Nasser's Egypt. The security repercussions of the coup and the Egyptian intervention posed a serious threat to the stability of the Saudi regime. In large part what made this threat particularly dangerous were Egypt's military capability and Yemen's geographic proximity. Yemen sat on the southern border of Saudi Arabia, sharing a frontier of 1458 kilometers. At the same time, Egypt had a large and well-equipped and trained army.

Given these two conditions, as Safran observed, the coup and the Egyptian intervention posed a threefold threat: first, an openly hostile republican regime backed by Nasser would generate a permanent security challenge to the Saudi regime; second, the presence of the Egyptian army in Yemen would empower the Saudi opposition, paving the way for a *coup d'état* against the Saudi royal family; third, an Egyptian invasion of the kingdom. Such a threat could have materialized under the pretext of a response to Saudi provocation, but in reality its intention was to bring about a decisive defeat of the Saudi forces, in order to encourage dissidents and separatists to rise against the royal family.[31] In fact, on several occasions, the Egyptian air force and navy bombed small Saudi villages near the Saudi-Yemeni border. Yet, despite the regularity of the Egyptian attacks the Saudi military retaliation did not come through. The heightened insecurity, however, compelled King Faisal to buy combat jets and a battery of Thunderbird surface-to-air missiles.

Responses: Low Internal Validation/Balancing, and External Balancing

The policy of appeasement changed drastically in light of the Yemeni coup and Egyptian military intervention on the side of the republicans against the royalists. Having failed to keep Nasser's threat at bay, Faisal abandoned appeasement and instead devised a defense policy based on a combination of internal validation, internal balancing, and external balancing. The additional income from oil revenue allowed Faisal to improve internal balancing and to some extent engage in internal validation.[32] At home, Faisal sought to boost support

31 Safran, *Saudi Arabia: the Ceaseless Quest for Security*, 94.

32 The extra cash was due to higher exports, but also as a consequence of an agreement between OPEC and oil companies: Saudi Arabia signed this agreement in 1964, gradually

for his regime with the ten-point program. The program stipulated legal and social reform and economic development. As noted above in the section of internal validation, the government expanded social welfare services and public employment.

Internal balancing efforts rested primarily on increasing financial spending on military and security apparatuses. Although spending on the military went up, it was the internal security agencies, which consumed the largest share of the budget, reflecting Faisal's anxiety about internal subversion. In the 1962–1967 period budget allocations to the Ministry of Interior climbed from 183 million SR to 3,859 million SR, and to the National Guard increased from 61 million SR to 202 million SR. In total spending on internal security apparatuses multiplied from 260 million SR in 1962 to 4202 million SR in 1967 (see Table 3.8). By 1967 internal security assumed 70 percent of total military spending.[33]

On the international front, the Saudi regime sought to enhance security through external balancing. Its balancing strategy proceeded in various directions and with different partners. First, the Saudi regime balanced with royalists against the republicans. It was a delicate balancing act, which aimed to frustrate Nasser's adventure in Yemen, but not to provoke him to take the war into the kingdom. Thus, the royalists received from Saudi Arabia financial aid and arms, but Saudi armed forces did not participate in the fighting.

Second, to counter Nasser's socialism and pan-Arabism, Faisal promoted an alignment based on Islamic ideology. To develop an alliance among Muslim states, Faisal invited Arab and non-Arab Muslim countries to join an international forum called the Islamic Conference. His efforts, however, did not materialize until 1969, at an Islamic summit in Rabat, which established the Organization of Islamic Conference (OIC). By then the threat of Nasser had vanished after the crushing defeat of the 1967 Six-Day War.

Third, Faisal sought British and American support for resisting Egyptian armed forces.[34] The response of the United States government was ambiguous. On the one hand, President Kennedy gave Faisal assurances regarding the territorial integrity and sovereignty of the kingdom; on the other hand, the

---

increasing the allowance of the oil-producing countries starting in January 1964 with an extra 3.5 cents a barrel.

33 In addition to internal validation and internal balancing, Faisal took the additional step of weakening the influence of pan-Arabism among the senior members of the ruling class by dismissing liberal princes and progressive commoners from the Council of Ministers. A group of loyal princes were appointed, including Prince Khaled as Deputy as Prime Minister, Prince Fahd as Minister of the Interior, and Prince Sultan as Minister of Defense.

34 Saudi Arabia broke off diplomatic relations with Britain in 1956. Relationship between the two countries resumed in 1963.

Kennedy Administration officially recognized the new Republican regime of North Yemen against the wishes of Saudi Arabia.[35] In spite of American oil interest in the kingdom, Faisal was unable to persuade Kennedy to withdraw his support to Nasser. Rather than confronting Egypt, the United States resorted to a relationship of cooperation based on economic aid as guarantees for safeguarding American interest in the region.[36] Having failed to win unequivocal American support, Faisal turned to the British, who were far more sensitive to Saudi Arabia's security needs.[37] The British were forthcoming with supplying arms and training to Saudi Arabia, and providing supplies to the royalists in North Yemen.[38]

Case Four: The 1967 Six-Day War

In the midst of the Saudi-Egypt confrontation in Yemen, tension escalated between the Arab states and Israel. It finally exploded in the early morning on June 5 when Israel launched its air attack on Egyptian and Jordanian air forces. The eruption of the 1967 War came at the height of Arab nationalism and a widespread anti-Western sentiment. With the outbreak of the war the situation inside the kingdom and throughout the region became highly volatile.

> The International Confederation of Arab Trade Unions Called on all Arab oil states to join in the embargo. It also urged workers to blow up

---

35 For more information on Egyptian-Saudi conflict and the Kennedy Administration, see Fawaz A. Gerges, "The Kennedy Administration and the Egypt-Saud Conflict in Yemen: Co-opting Arab Nationalism," *Middle East Journal* 49 (Spring 1995): 292–311.

36 With the arrival of President Johnson in November 1964, US-Egypt relations began to deteriorate. Unlike Kennedy, President Johnson was more responsive to Faisal's requests in the region aimed at subduing Nasser, for several reasons. Johnson did not approve of Nasser's nationalism, or his rising prestige and influence in the region. Moreover, Johnson believed Arab nationalism had been exploited by the Soviet Union to weaken the interest of the West in the Middle East. See William B. Quandt, *Decade of Decisions* (California: California University Press, 1977), 38.

37 There were several reasons behind Britain's cooperation with Saudi Arabia. First, the republican regime in North Yemen posed a direct threat to the British position in Aden and South Yemen. Second, Britain and Nasser's Egypt were on bad terms since the Suez Crisis, which elevated the status of Nasser and humiliated the British. Yemen was a payback opportunity. Third, Nasser's expansion on the Arabian Peninsula constituted a threat to the security of oil supplies.

38 For a discussion on the British involvement and the relationship between North and South Yemen, see Fred Halliday, *Arabia without Sultans*, reprint, 1975 (Middlesex, England: Penguin Books, 1974), 188–222.

> pipelines and oil installations in any Arab nation that refused to comply with the embargo.[39]

To make matters worse, Nasser accused the United States and Britain of providing air cover support to Israel against Egypt and Jordan. The charges inflamed an already high anti-Western sentiment throughout the Arab world, and it created a major security problem for Saudi Arabia and other Arab oil-producing countries. In Saudi Arabia, the security situation became acutely alarming. As noted in the section above, pan-Arabism was a major force of disruption and opposition to the regime. During the 1950s and 1960s there were frequent protests, especially in the months leading to the 1967 War. Two days after the war erupted, riots broke out in Dharhan. The protesters stormed ARMACO's office and attacked the American Consulate. In Riyadh the National Guard had to intervene in order to break up a demonstration of several thousand people.

Responses: External Validation

One day after Israel launched a surprise attack against Egypt and Jordan, the Arab oil-producing countries enacted an oil embargo against the United States and Britain for their alleged support of Israel. Reluctantly the Saudi regime joined the embargo in order to avoid domestic instability and isolation in the Arab world. The embargo did not unfold in a united and coherent form; instead Arab oil-producing countries followed different courses of actions. For example, Iraq imposed a total embargo on oil exports for a few weeks, while Gulf countries including Saudi Arabia placed an embargo for about a week. At the end of the week, Saudi Arabia continued the ban on shipments only to France, Britain, and West Germany.[40]

The decision of Saudi Arabia to stand by Egypt in times of war seemed paradoxical, given, on one hand, Nasser's hostility to the Saudi royal family, and on the other hand, the strategic economic-security alliance with the United States. In the end, the Saudi royal family concluded that unless the highly inflamed

---

39 M.S. Daoudi and M.S. Dajani, "The 1967 Oil Embargo Revisited," *Journal of Palestine Studies*, no. 13 (Winter 1984), 69.

40 Like the 1956 oil embargo, the 1967 embargo did not lead to dramatic political and economic impact on targeted countries. There were many reasons for its ineffectiveness. M.S. Daoudi and M.S. Dajani list ten reasons. But perhaps the most important are the availability of stock, availability of tanker transport, availability of alternative sources (such as the United States, Iran, and Venezuela), and the short duration of the embargo. See M.S. Daoudi and M.S. Dajani, "The 1967 Oil Embargo Revisited," 8–86.

public opinion at home and throughout the Arab region was accommodated, they would risk confronting a mass uprising.

Like the 1956 embargo, the 1967 embargo did not produce an energy crisis. Supplies from other oil-producing countries were readily available to offset the missing 1.5 mbd. Instead, the embargo created a major challenge for international oil companies to devise supply schemes in which non-Arab oil could be diverted to the three embargoed countries, the United States, Britain, and Germany; while Arab oil originally destined for embargoed countries was diverted to other countries. Oil shortages were compensated for by increasing production from other oil-producing countries, namely the United States, Iran, Indonesia, and Venezuela. By the end of June Saudi Arabia was anxious to resume oil shipments. Saudi Oil Minister Sheikh Ahmed Zaki al-Yamani took on the lead in calling for an end to the embargo. Yamani urged the Arab leaders to consider the negative consequences of the embargo on their economies. King Faisal expressed similar views. In an official statement on July 7, 1967, the Saudi government declared:

> Now that it has been established that there was no evidence of British and American aircraft helping Israel in last month's war, there is no reason to continue the ban on exporting oil to the two countries.[41]

However, the Saudi effort was defeated by strong opposition from the other Arab governments, who "accused the Saudis of "deserting the Arab cause" and "undermining Arab solidarity.""[42] Under these circumstances, out of the necessity of security, the regime was obliged to reaffirm its Arab credentials by pledging to maintain the boycott of shipments to the United States and Britain. The embargo lingered on until the fourth Arab Summit of August 29, 1967, when the 12 nations agreed to resume full production.

Conclusion

The confluence of major growth in world demand for oil and the quadrupled rise of Saudi oil exports from 346 million barrels in 1950 to 1178 million barrels in 1969 made the kingdom one of the most important oil producers in the world. The oil concession, however, limited the extent to which the Saudi government could employ oil resources for security purposes. With ARAMCO in

41 Quoted in M.S. Daoudi and M.S. Dajani, "The 1967 Oil Embargo Revisited," 72.

42 Ibid.

full control of decision-making for oil pricing and supply, the Saudi state was merely a recipient of oil allowance calculated on the basis of royalties and profit-sharing. This oil income accounted for the majority of total state income. At the same time, despite the growing demand for oil there were plenty of supplies of oil in the market.

With respect to validation, the Saudi state did embark on rapid development and financed major infrastructure and development projects. It also set up public corporations that dominated various sectors of the economy including industrialization, finance, and services. As institutions proliferated, a vast amount of employment opportunities opened up in the civil administration and the public sector. Similarly, military and security apparatuses were built, from the ground up, into professional organizations equipped with modern arms, technology, and training. New repression agencies were set up with directives of monitoring, controlling, and enforcing internal security and stability. In effect, the growth of penetrative, pervasive bureaucratic machinery enabled the state to engage in distribution of rewards (internal validation) and coercion (internal balancing). However, spending was constrained by low prices and profit-sharing arrangements between the Saudi state and ARAMCO.

In the international arena, the Saudi regime used external validation during the 1956 Suez Crisis and the 1967 Six-Day War. In both events, the Saudi regime reluctantly participated in an oil embargo against Western countries in order to accommodate public sentiment at home and in neighboring Arab countries. Both embargoes, however, were ineffective and short-lived. Supplies from non-Arab oil-producing countries compensated for shortages. In the case of the 1956 embargo, most of the extra supply came from the United States. In 1967, oil supplies increased from the United States, Iran, Venezuela, and Indonesia.

In response to Nasser's pan-Arabism, Faisal employed internal validation and internal balancing at low intensity. Abroad, however, the Saudi regime attempted external balancing. In order to prevent Nasser from securing his grip on Yemen, the Saudi regime provided financial aid and arms to the royalists. This balancing, however, did not prevent Nasser from bombing Saudi border villages. Furthermore, Faisal sought to distant the Kennedy Administration from Nasser. The last balancing act Faisal pursued was to form an alliance among Islamic states to counter the spread of Nasser's influence in the Arab world and other Islamic countries.

CHAPTER 4

The First Boom (1970–1985)

Two important structural changes in the oil market took place in the early 1970s, which dramatically and positively influenced the capabilities of the Saudi Arabian state in the domestic and international arenas. First was the depletion of oil surpluses in other oil-producing countries – mainly the United States, which had been one of the most important suppliers of oil in the non-communist world. Second was the transition of control over oil resources from oil companies to producing states. As a result of these two factors, the Saudi state acquired unprecedented economic capabilities. This chapter explores the effects of these changes on the Saudi's state security strategies during the boom. The chapter is divided into three sections. Section one examines global oil supply and demand of oil in the market as well as Saudi Arabia's oil exports and proven reserves. While section two deals with strategies of internal balancing and validation, section three considers responses of the Saudi state to four specific threats.

Pattern and Domain of Dependence in the Boom Era (1970–1985)

Oil Market: Supply and Demand

Although during the 1970s the postwar economic boom had lost some of its momentum, in the 1973–1980 period the world industrial output had a modest average growth of 2.4 percent.[1] Demand for oil therefore continued to climb, despite the two major oil shocks of 1978 and 1978–79. Table 4.1 shows that oil consumption dipped in 1974 in the wake of price surges of October 1973 and January 1974, but rebounded in 1976 and remained strong until 1978. The second shock came on the heels of the Iranian revolution and the outbreak of the Iran-Iraq War in 1980. As a result of these two events the price of oil reached $17.26 a barrel, prompting lower consumption in OECD countries after 1979. Consumption in OECD countries decreased from 44 mbd in 1978 to 37 mbd (15 percent) in 1985. However, over the same period, consumption increased in non-OECD countries. Overall, weak demand caused the world's consumption to decrease in the 1978–1985 period by 6.3 percent.

1 Kennedy, *The Rise and Fall of the Great Powers: Economic Change and Military Conflict from 1500 to 2000*, 414.

 | DOI 10.1163/9789004277731_006

TABLE 4.1 *World oil consumption (Million b/d), 1970–1985*

Year	Total (OECD)	Total Non-OECD	World
1970	35	12	47
1972	39	14	53
1974	40	16	57
1976	42	18	60
1978	44	20	64
1980	42	21	63
1982	38	22	60
1984	38	22	60
1985	37	23	60

SOURCE: EIA, ANNUAL ENERGY REVIEW 2007.

On the supply side, production was increased in many regions. Between 1970 and 1981, surplus capacity in the world was lower than in the 1960s due to strong demand for oil, which required many producing countries to maximize output. Between 1970 and 1978 the world production increased to 24 percent. As shown in Table 4.2, except for North America and Latin America, production increased in the rest of the world. As a result of shortages from two major oil producers, Iran and Iraq, excess capacity drastically declined in the 1979–1981 period. Without sufficient supplies to compensate for the missing oil, the price of oil surged. Given that there were few alternative supplies in the world, the balance of power rested on the side of the producers.

After 1980 there was a shift from surplus to glut in the market. In response to the lower world demand for oil, production decreased from 59 billion barrels in 1980 to 52 billion barrels in 1985 (12 percent). Much of this reduction came from non-OPEC countries. In order to shore up prices, between 1980 and 1985, OPEC countries (excluding Saudi Arabia) decreased output from 13 billion to 12 billion barrels (4 percent), and Africa decreased output from about 6 billion to about 5 billion barrels (20 percent). The largest cutback, however, came from Saudi Arabia. As a swing producer Saudi Arabia dropped production from almost 10 billion to 3 billion barrels (68 percent). On the other hand, production went up in non-OPEC countries. Production increased in North America by 3.3 percent, in Latin American by 9.4 percent, in Western Europe by 53 percent, and in Asia and the Pacific by 12.5 percent.

TABLE 4.2 *Oil production by region (1000 b/d), 1970–1985*

Year	North America	Latin America	Eastern Europe	Western Europe	Saudi Arabia	Africa	Asia and Pacific	Middle East[1]
1970	10,963	5,175	7,235	470	3,799	6,032	1,544	9,980
1972	11,061	4,838	8,028	463	6,016	5,684	2,382	11,925
1974	10,409	4,801	9,335	486	8,480	5,382	3,298	13,226
1976	9,387	4,340	10,491	921	8,577	5,894	4,111	13,470
1978	9,957	4,765	11,525	1,792	8,301	6,066	4,726	12,822
1980	9,891	5,575	12,088	2,519	9,901	6,062	4,917	8,445
1982	9,776	6,238	12,329	2,989	6,483	4,401	4,623	6,447
1984	10,149	6,174	12,300	3,697	4,079	4,606	5,251	6,439
1985	10,222	6,098	11,986	3,853	3,175	4,868	5,532	6,550

(1) Saudi Arabia is not included.

SOURCE: OPEC, ANNUAL STATISTICAL BULLETIN, 2007.

Saudi Arabia: Oil Reserves and Exports

Besides the tight oil market in the1970–1981 period, several other factors turned this era into a golden age from the perspective of Saudi Arabia. First and foremost was the transfer of power over oil resources from ARAMCO to the Saudi state. By early 1974 the Saudi state was in full control of exploration and production (i.e., upstream), and for the first time in full charge of issues pertaining to exports and prices. Second, the combination of vast reserves and excess productive capacity allowed the kingdom to play a crucial role in effecting oil price and supply. Inside and outside of OPEC, the weight of Saudi Arabia as an oil superpower was quickly recognized. Its oil reserves were by far the largest in OPEC, and constituted more than one-third of total OPEC reserves. No other country or region had a comparable amount of proven reserves (see Table 4.3).

The kingdom was both a top producer and exporter. Between 1970 and 1981, Saudi Arabia's production averaged 38 percent of total OPEC production. No other country within OPEC had similar export capacity. As Table 4.4 shows, in 1974 the kingdom exported roughly 8 mbd, that is, about 3 million more than the second largest exporter, Iran, and about 4 million more than the third largest exporter, Iraq. Third, in the non-communist world the most important oil producer, the United States, had turned into a major net importer. Despite its high production volume of 8 mbd to 9 mbd, consumption in the United

TABLE 4.3 *Proven oil reserves by region as percentage of total world reserves, 1970–1985*

Region/Country	1970	1975	1980	1985
North America	9.1	6.2	5.7	4.5
Latin America	4.8	5.7	11.5	15.7
Eastern Europe	11.1	9.7	10.4	8.5
Western Europe	1.3	3.1	2.5	1.9
Saudi Arabia	25.8	26.8	26.1	22.7
Africa	9.3	9.3	8.2	7.4
Asia and Pacific	3.1	5.2	5.2	4.9
Middle East[1]	35.5	34	30.3	34.4

SOURCE: OPEC, ANNUAL STATISTICAL BULLETIN, 2007.

TABLE 4.4 *Oil exports by region (1000 b/d), 1970–1985*

Year	North America	Latin America	Eastern Europe	Western Europe	Saudi Arabia	Africa	Asia and Pacific	Middle East[1]
1970	676	2,572	1,323	42	3,217	5,852	859	8,926
1972	865	2,350	1,517	111	5,444	5,401	1,143	10,918
1974	819	2,115	1,587	132	7,922	5,081	1,403	12,053
1976	413	1,858	2,196	432	8,032	5,398	1,760	12,237
1978	404	1,973	2,280	839	7,706	5,357	1,933	11,376
1980	475	2,515	2,415	1,254	9,223	5,380	1,731	7,153
1982	441	3,017	2,430	1,628	5,639	3,235	1,643	4,770
1984	534	3,027	2,563	2,182	3,187	3,477	2,000	4,891
1985	687	2,791	2,299	2,294	2,151	3,841	1,985	4,941

(1) Saudi Arabia is not included.

SOURCE: OPEC, ANNUAL STATISTICAL BULLETIN, 2007.

States was in the range of 15 mbd to 18 mbd. To make up for the difference, the United States imported somewhere from 6 mbd to 8 mbd. The Soviet Union was also a large oil producer, but much of its output was locally consumed. Some of its oil supplies were exported to communist countries, such as Eastern Europe, North Korea, and Vietnam, as well as other non-communist countries such as India, Bangladesh, and Sri Lanka.[2] On the other hand, between 1970 and 1985 Saudi Arabia exported somewhere between 67 percent and 93 percent of its output.

Patterns of Dependence during 1970–1985

The 1970 to 1985 period reflects two types of dependence, one between Saudi Arabia and the consumers (1970–1981), and, second, between Saudi Arabia and the other oil rentier states (1981–1985). In the first period, consumers' opportunity costs were high, while producers' opportunity costs were low, because the market was driven by demand. When consumers are dependent on the rentier state, it is expected that the rentier state is likely to employ internal balancing and validation at high intensity and have the means to employ external validation and external balancing. When the market slipped from excess demand to excess supply, relations with the consumers began a gradual shift toward asymmetric dependence. At the same time, during the period 1981–1985, oil price was under pressure of weak demand and expanding supply. Because Saudi Arabia had the excess capacity to push prices up or down, its relationship with other rentier states turned to asymmetric dependence. Consequently, Saudi Arabia was well positioned to formulate strategies of external validation and external balancing.

Overall, in this period the economic position of the kingdom substantially improved, due to the big jump in oil price. As shown in Table 4.5, revenues almost tripled in 1974 when oil prices went up in just one year from $2.70 a barrel to $9.76 a barrel. The second largest jump came in 1979 and 1980 as oil prices more than doubled from $12.70 in 1978 to $28.67 a barrel in 1980. As a result of growth in oil revenues, the Saudi state experienced budget surplus in eleven years out of the sixteen years. The huge amount of wealth transfer from consuming countries substantially enhanced the financial capability of the Saudi regime and facilitated greater spending on strategies of internal validation and internal balancing. These strategies are discussed in the next section.

Internal Validation and Internal Balancing in the Boom Era

Internal Validation

The influx of enormous wealth in the 1970s and the early 1980s left the Saudi state with huge surpluses and the challenge of coming up with grand schemes

2 Arthur Jay Klinghoffer, "SinoSoviet Relations and the Politics of Oil," *Asian Survey*, no. 16 (January 1976), 548.

TABLE 4.5 *Saudi Arabia revenue, and budgetary deficits and surplus, 1970–1980*

Fiscal year	Oil revenue (Million SR)	Oil revenue as percentage of total revenue	Oil price arabian light	Deficit/surplus (Million SR)
1970–71	7,122	90	1.3	1,647
1971–72	9,685	87	1.7	2,990
1972–73	13,480	88	1.9	5,210
1973–74	39,285	94	2.7	23,110
1974–75	94,190	94	9.8	65,064
1975–76	93,481	90	10.7	22,149
1976–77	1,21,191	89	11.5	7,684
1977–78	1,14,042	87	12.4	-7,389
1978–79	1,15,078	88	12.7	-16,467
1979–80	1,89,295	90	17.3	25,472
1980–81	3,19,305	92	28.7	1,11,345
1981–82	3,28,594	89	34.2	83,356
1982–83	1,86,006	76	31.7	1,270
1983–84	1,45,123	70	28.8	-23,767
1984–85	1,21,348	71	28.1	-44,854
1985–86	88,425	66	27.5	-50,439

SOURCE: SAUDI ARABIAN MONETARY AGENCY (SAMA), ANNUAL REPORTS, VARIOUS YEARS.

of wealth distribution, which would boost the image of the royal family. It initiated a series of five-year development plans. These plans placed the government in the driver's position behind all economic transactions. The primary objective of these plans was wholesale development of the physical infrastructure and improvement of the social and economic welfare of the population, along with efforts to ensure security and stability. In the First Development Plan it was proclaimed that,

> The general objectives of economic and social development policy for Saudi Arabia are to maintain its religious and moral values, and to raise the living standards and welfare of its people, while providing for national security and maintaining economic and social stability.[3]

3 The Kingdom of Saudi Arabia, Ministry of Economy and Planning, *First Development Plan 1970–1975*, 23. Accessed March 2007. http://www.planning.gov.sa.

TABLE 4.6 *Total government expenditure on development (Billion SR)*

Sector	First plan	Second plan	Third plan
Economic development[1]	3	92	262
Human development[2]	7	80	130
Social development[3]	2	33	61
Physical infrastructure[4]	12	113	249
Administration	8	38	31
Emergency & subsidies	–	63	50
Total	32	419	783

(1) Includes projects related to water, agriculture, industry, trade, services
(2) Includes projects related to the general education system.
(3) Includes projects related to health, social security, social rehabilitation, etc.
(4) Includes projects related to transportation, telecommunications, and housing.
SOURCE: MINISTRY OF ECONOMY AND PLANNING, FIRST DEVELOPMENT PLAN (1970–75), 43; SECOND DEVELOPMENT PLAN (1975–80), 531; THIRD DEVELOPMENT PLAN (1980–85), 88.

Similarly, the goals of the Second and Third Plans were declared to be: (1) to uphold the values of Islam, by applying and spreading the sharia, (2) to defend Islam and the kingdom, and to ensure internal security and stability, (3) to maintain economic growth by extending the time horizon of oil revenues, and (4) to diversify away from oil as the primary source of national income.

Because of the financial size, manpower requirements, and scale of construction the development plans influenced all sectors of the economy. The data in Table 4.6 show the broad scope and massive spending of the First, Second, and Third Development Plans. Unlike subsequent plans, the First Plan was formulated under financial constraints. It was initiated only one year after the government had recovered from a deficit of 360 million SR. Thus, spending in the First Plan was relatively modest in comparison with the Second and Third Plans. Subsequent Second and Third Development Plans were devised under more favorable economic conditions, and thus the government drastically increased allocations to development projects and welfare programs. Spending in the First Plan was outpaced by a twelve-fold increase in the Second Plan and by a twenty-four-fold increase in the Third Plan. As a direct result of these plans the standard of living greatly improved, because of generous subsidies on a wide range of goods and services, dramatic growth in employment opportunities, and the stimulation of the private sector.

TABLE 4.7 *Growth in civilian labor force (1000)*

Sector	1966	1970	1975	1980	1985
Producing sectors[1]	628	799	1,072	1,424	2,067
Service sectors[2]	253	488	1,042	1,158	1,845
Government	110	184	321	399	469
Oil	15	22	36	45	65
Total	1,006	1,493	2,471	3,026	4,446

(1) Includes sectors related to agriculture, mining, construction, utilities, etc.
(2) Includes sectors related to trade, finance, transport, and other services.
SOURCE: MINISTRY OF ECONOMY AND PLANNING, FIRST DEVELOPMENT PLAN (1970–75), 81; SECOND DEVELOPMENT PLAN (1975–80), 68; AND THIRD DEVELOPMENT PLAN (1980–85), 93.

In terms of public services, the government expanded education and health programs. The benefits provided during the Second Plan constituted almost an extra 29 percent contribution to personal income.[4] Among the four development sectors, however, the bulk of allocation went to infrastructure development. In fact, the share of social and human development declined in the Second and Third plans. The data in Table 4.6 show that allocations to human development declined by 2 percent in the Second Plan, while human development declined by 5 percent in the Third Plan.

Employment experienced dramatic growth. Implementation of the development plans created strong demand for manpower in almost all sectors and for all types of skills. The requirements of the Second and Third Plans in particular, however, exceeded the indigenous labor force. Table 4.7 provides an indication of the scale of growth in employment in five sectors of the economy. All sectors experienced growth in employment, but the largest growth was in the producing sector, mainly because construction projects demanded a lot of manpower. While in the Second Plan employment went up by 555,000, in the Third Plan total employment increased by 1,420,000 jobs. A large portion of this increase, however, came from foreign labor. Unable to fully meet the growth in employment from the indigenous workforce, the Saudi regime responded by adopting two policies: introducing training programs designed to improve the skills of the Saudi citizens, and importing foreign labor. In 1980,

4 The Kingdom of Saudi Arabia, Ministry of Economy and Planning, *Third Development Plan 1985–1990*, 15. Accessed March 2007. http://www.planning.gov.sa.

TABLE 4.8 *Government subsidies (Million SR), 1970–1985*

Year	Food	Agricultural	Social security	Electrical	Total
1970	–	–	39	1	40
1971	–	–	45	1	46
1972	–	4	55	15	74
1973	300	20	76	14	410
1974	750	69	358	13	1,190
1975	700	333	500	25	1,558
1976	600	603	653	175	2,031
1977	700	772	864	312	2,648
1978	800	829	971	563	3,163
1979	1,450	586	1,005	749	3,790
1980	3,000	766	986	1,234	5,986
1981	5,000	1,129	1,391	2,586	10,106
1982	4,150	1,472	1,586	2,816	10,024
1983	1,600	1,173	1,540	3,548	7,861
1984	2,633	1,478	1,485	2,750	8,346
1985	1,950	994	1,461	1,750	6,155

SOURCE: ADOPTED FROM CHAUDHRY, 1997, 150.

out of the total workforce of 3,026,000, about 1,493,000 were Saudi. Five years later this number modestly increased to 1,786,000.

The private sector became heavily dependent on government spending. In almost all sectors the government outsourced the implementation of its planned projects to the private sector. Thus, the performance of the private sector became largely driven by the size of government expenditure. The extent of this dependence is illustrated in the ratio of autonomous activities to non-oil GDP during the Second Plan. As a result of growth in the volume of government contracts in the first year (1976) this ratio declined from 54 percent to 35 percent. It rebounded to 48 percent in 1999, but then declined again to under 40 percent in 1980.[5]

Subsidies on a variety of consumer goods and services also dramatically increased after 1973. In Table 4.8 estimates of subsidies are listed for each sector. The last column shows the magnitude of expansion. As a percentage of oil revenues subsidies have varied from a low of 1.1 percent in 1973 to a high of 6.8

5 Ibid., 32.

percent in 1984. However, the figures in Table 4.8 represent an approximate value of the subsidies and only a fraction of the cost involved for the state. It does not, for instance, include fuel, housing, press, industry, and water subsidies. To appreciate the full magnitude of the costs for the state, it is also necessary to distinguish between non-capital and capital subsidies.[6] The non-capital subsidies target current production and consumption activities, such as the items listed in Table 4.9, while capital subsidies refer to public loans advanced at subsidized interest, such as items listed in Table 4.9. The figures in Table 4.9 show the amount of credit disbursed to the private sector and public corporations. The figures, however, are incomplete because they do not take into account the opportunity cost of not investing the wealth allocated to subsidies. As Askari points out, "From the perspective of the government, if capital subsidies are not given, these funds can be invested, resulting in revenues (interest) and the repayment of principle at maturity."[7] Nevertheless, summing both types of subsidies gives an approximation of the magnitude of state spending. With the government situated as both the receiver and distributor of oil wealth, state-society relations were altered. First, the society became fully dependent on the state for its welfare, including receiving free health and education and guarantees for employment. Although the economic benefits were not equally distributed, everybody was better off. In exchange for the generous material benefits, the government required little of its citizens – mainly political passivity. The state demanded neither taxation nor conscription. Second, political activism became chiefly concerned with economic gains. The state became a lucrative arena for individuals and groups to serve their economic interests by competing for government contracts, commissions, and employment. As Luciani points out:

> To the individual who feels his benefits are not enough, the solution of manoeuvring for personal advantage within the existing setup is always superior to seeking an alliance with others in similar conditions.[8]

Thus, the citizens were led to believe that they have less to gain and more to lose by opposing the state. In sum, the royal family sought to create a domestic environment in which the society at large was economically rewarded for political acquiescence.

---

6 Hossein Askari, *Saudi Arabia's Economy: Oil and the Search for Economic Development* (London: Jai Press, In., 1990), 102–120.

7 Ibid., 103.

8 Luciani, "Allocation vs. Production States," 74.

TABLE 4.9 *Subsidized government loans (Million SR), 1969–1985*

Year	Saudi Agricultural Bank[1]	Saudi Credit Bank	Specialist Funding Programs[2]	Public Investment Fund[3]	Saudi Industrial Development Fund[4]	Real Estate Development Fund	Total
1969	16	–	–	–	–	–	16
1970	17	–	–	–	–	–	17
1971	16	–	–	–	–	–	16
1972	20	–	2	134	–	–	156
1973	36	9	7	263	–	–	315
1974	145	40	48	603	35	–	871
1975	269	81	196	1,512	1,699	2,159	5,916
1976	490	158	433	3,843	2,274	8,901	16,099
1977	586	103	402	6,267	4,340	7,534	19,232
1978	709	55	592	3,893	7,657	5,766	18,672
1979	1,129	38	635	2,949	6,486	8,575	19,812
1980	2,531	372	236	6,032	6,660	7,598	23,429
1981	2,933	331	122	8,772	5,347	7,141	24,646
1982	4,166	243	184	9,452	5,116	8,307	27,468
1983	3,496	233	222	5,847	5,216	8,909	23,923
1984	2,322	252	301	3,662	2,416	8,598	17,551
1985	1,551	276	139	1,429	926	6,795	11,116

(1) Includes loans for working capital;
(2) Includes loans to contractors, Private hospitals and dispensaries, hotels, newspapers, bakeries etc.;
(3) Including equity investments;
(4) Includes industrial, electricity, cold storage and dates packing.

SOURCE: MINISTRY OF ECONOMY AND PLANNING, ACHIEVEMENTS PLAN, 2000.

Internal Balancing

The sharp rise in oil revenues also caused spending on security to rise. The massive allocations were mainly devoted to acquiring expensive military hardware, building new military facilities, and enlisting training programs. In 1970 the Saudi Defense Ministry commissioned the United States Department of Defense to carry out a survey of Saudi defense deficiency and to make recommendation on how to improve the strength of the armed forces. A second survey was commissioned in 1974. Following up on the recommendations of both surveys the Saudi government conducted many costly deals for improving the capabilities of the air force, navy, and armed forces. Similarly, the National Guard underwent a comprehensive modernization program consisting of new equipment, a training program, and new housing facilities. The price tag reached 1.9 billion dollars.[9] The modernization of the military and security services constituted the largest public expenditure of the budget. Between 1970 and 1985, total military and security allocation steadily increased from 2.6 billion SR to 81.3 billion SR (see Table 4.10). In terms of percentage of oil revenues, security expenses dramatically increased from 36 percent in 1970 to 67 percent in 1984. The allocations, however, were unevenly divided between internal and external security apparatuses. The percentage of total revenues

TABLE 4.10 *Defense and internal security allocations (Million SR), 1970–1985*

Security expenses	1970–1971	1972–1973	1974–1975	1976–1977	1978–1979	1980–1981	1984–1985
Ministry of defense	1,727	3,157	7,217	27,188	29,912	47,796	47,546
Ministry of information	76	165	321	1,165	1,136	1,442	1,198
National guards	282	404	1,295	4,669	5,393	7,509	11,482
Intelligence bureau	22	49	96	206	400	486	838
Ministry of interior	508	980	2,304	5,852	9,027	12,960	20,296
Total	2,615	4,755	11,233	39,080	45,868	70,193	81,360

SOURCE: MINISTRY OF ECONOMY AND PLANNING, ANNUAL STATISTICAL BOOKS 1970, 1975, AND 1980.

9 Safran, *Saudi Arabia: The Ceaseless Quest for Security*, 204–212.

allocated to internal security (Ministry of information, National Guards, Intelligence Bureau, and Ministry of Interior) averaged 35 percent a year, whereas percentage of revenues allocated to the Ministry of Defense averaged 65 percent a year.

Despite the massive increase on defense and security, however, the defense capabilities of the kingdom remained weaker than those of its two main adversaries, Iraq and Iran. The military weakness of the kingdom partly had to do with the size of the armed forces, which was dwarfed by that of Iraq and Iran. Instead of remedying the manpower shortage by introducing mandatory military service, the royal family opted to compensate for this shortage by buying advanced military equipment. The Saudi regime avoided military conscription for fear that it might produce unwanted consequences – for instance, citizens demanding political reforms and participation in the political process. A second likely reason is that a smaller armed force is more easily controlled, and therefore is more likely to be deterred from playing a role in an internal coup against the royal family. Consequently, the kingdom had no choice but to rely on external military support in the face of an imminent external threat.

Cases of Major Threats

The following cases reveal the strategies adopted by the Saudi state in response to immediate threats. Unlike the previous section, which examined the intensity of internal validation and internal balancing, this section explores the different types of strategies employed by the Saudi state in response to the October 1973 War, the Camp David Accords, the Iranian Revolution, and the Iran-Iraq War.

Case Five: The October 1973 War

As in the cases of the 1956 Suez Crisis and the 1967 Six-Day War, once again the Saudi regime was confronted with a war situation that posed a serious threat to its stability. The October 1973 War (or Yom Kippur War) broke out with a surprise attack by Egypt and Syria on Israel.[10] The surprise attack was unleashed

10 The war was part of an ongoing conflict between the Arab countries and Israel. The root of the conflict is the creation of the state of Israel in historical Palestine. See Ilan Pappe, *A History of Modern Palestine* (Cambridge: Cambridge University Press, 2004). The Saudis' solidarity with the Palestinian cause is grounded in two principles, Islam and Arabism. These two principles lie at the heart of the Saudi identity. The Islamic religion units the disparate Muslim states as one community (*umma*). Moreover, Palestine is the

on two fronts. In the north the Syrian troops attacked Israeli defense lines in the Golan Heights, whereas in the south the Egyptian troops advanced across the Suez Canal. In the initial days of the attack, Egypt and Syria benefited from the element of surprise. In the Sinai the Egyptian troops secured a position seven to ten kilometers east of the Canal. Similarly, Syrian attack inflicted heavy losses on the Israeli troops.

However, by October 9 the tide of war shifted in favor of Israel. On October 9–10 the Israeli armed forces managed to halt the advancement of the Egyptian troops in the Sinai. On the Golan front, Israel steadfastly responded to the Syrian attacks, and on October 10 Israel launched a major offensive attack against Syrian positions. To help ease the Israeli pressure on Syria, on October 14 the Egyptian Second and Third Armies launched a major offensive. Not only was the Egyptian offensive a failure, but Israel also went on the offensive. To make matters worse for the Arab armies, the United States begun an airlift to re-supply Israel with arms on October 13. Moreover, on October 19, Congress approved 2.2 billion dollars' worth of military assistance to Israel. On October 15, under the command of General Sharon, Israeli forces crossed the Canal with the intention of encircling the Egyptian Third Army.[11] By October 20 Israel was clearly on the offensive, and a third Arab defeat was on the way.

The ominous outcome of the war and the American intervention on the side of Israel left Saudi Arabia in a vulnerable position. As Safran rightly concluded, the danger of the 1973 War presented the Saudi regime with two types of security concerns. A defeated Egypt could estrange President Sadat, turning Egypt away from the Western bloc toward the Soviet Union for military aid and political support. Such a scenario could lead to increase in the

home of the third holiest place in Islam, the Sacred Noble Sanctuary (Al-Aqsa Mosque and Dome of the Rock) or *al-Haram al-Sharīf*. Because of its Wahhabi tradition and custody of the two holiest places, Mecca and Medina, Saudi Arabia considers itself the center of the Muslim world and responsible for Islamic causes. Arabism also fuels Saudi support for the Palestinians. The Saudi society considers Palestine as an integral part of a larger Arab nation. Noble observes that, "The population of the area has a sense of Arab identity that transcends individual nationalities and encourages a sense of kinship." See Paul C. Noble, "The Arab System: Opportunities, Constraints, and Pressures," in Bahgat Korany and Ali E. Hillal Dessouki, eds., *The Foreign Policies of Arab States* (Colorado: Westview Press, 1984), 45.

11 Kenneth M. Pollack, *Arabs at War: Military Effectiveness, 1948–1991* (University of Nebraska Press, 2007), 106–122.

presence and influence of the Soviets in the region.[12] Alternatively, Sadat might be ousted, and possibly replaced "by a leadership of unpredictable orientation or even by chaos."[13] Either way, a third Arab defeat, King Faisal reasoned, would endanger his dynasty.[14] Moreover, in the wake of the United States military support to Israel, the position of the Saudi regime became more precarious. As an oil supplier to the United States, the regime was a prime target of angry public opinion at home and in neighboring Arab countries.

Responses: External Validation

The oil embargo of 1973 exemplified the usage of oil resources as an external security strategy. Withholding access to oil resources was transformed into a mechanism of punishing and rewarding individual countries in order to ensure domestic and regional stability. Once again the oil weapon was unleashed in response to the Arab-Israeli conflict. Unlike the previous embargoes, the 1973 oil embargo was more sophisticated and influential. The effectiveness of the embargo lay in its discriminatory, flexible, and gradual approach. Oil-consuming countries were divided into four categories, and given the possibility of avoiding the boycott if a pro-Arab position was adopted. The categories were: most favored, preferred, neutral, and embargoed. The first category comprised countries that were allowed to import their full demand for Arab oil; the second category comprised countries that were allowed to import either the average of the first nine months of 1973 or an amount equal to the amount imported during September 1973; the third category comprised countries whose imports were reduced; the fourth category comprised countries whose supplies of Arab oil were fully discontinued, namely the United States and the Netherlands, but subsequently extended to include Portugal, South Africa, and Rhodesia.[15] The gradual approach adopted by Saudi Arabia within the Organization of Arab Petroleum Exporting Countries (OAPEC) reflected Saudi Arabia's intention to pressure, not to completely break off with the United States. As Table 4.11 shows, cutbacks were gradual,

12 After the Six-Day War, Israel became a regular recipient of American aid, while Egypt enlisted the help of the Soviet Union for arms transfer and military training. The Saudis were deeply concerned about the presence of the Soviet Military Mission in Egypt and feared that another devastating defeat would only increase the influence of the Soviet on Egypt.

13 Safran, *Saudi Arabia: The Ceaseless Quest for Security*, 154.

14 Holden and Johns, *The House of Saud*, 337–338.

15 Seymour, *OPEC: Instrument of Change*, 116–122.

TABLE 4.11 *The Arab oil embargo of 1973.*

Date	Events
17-Oct-73	Arab oil produces announced 5% monthly cutback of oil to the US and other pro-Israel nations, using previous month's figure as basis for cutbacks.
18-Oct-73	Saudi Arabia announced 10% cutback in oil production, threatening further cutbacks if the US continues to support Israel.
20-Oct-73	Saudi Arabia decided to halt all oil exports to the US because of its support to Israel
23-Oct-73	Embargo was extended to the Netherlands
04-Nov-73	Arab producers increased the cutback in oil production to 25% below the September 1973 level. In actuality, Saudi Arabia's output reached 35% below the daily average of September.
18-Nov-73	Arab producers canceled the 5% cutback for Western Europe, which was scheduled for December.
23-Nov-73	The embargo extended to Portugal, Rhodesia, South Africa
09-Dec-73	Arab producers announced additional 5% cut in January for non-friendly countries
25-Dec-73	Arab oil producers canceled January 5% cutback. Saudi Arabia announced 10% increase in output.
18-Mar-74	Arab oil producers ended the embargo to the United States

turning the wheel of pressure on a monthly basis to produce a pro-Arab stand.

At the height of the embargo during the months of November and December, the Arab oil production was reduced by approximately 5 mbd. As shown in Table 4.12, Saudi Arabia alone was responsible for the removal of more than 2 mbd in November.[16] By December the international oil market had lost 4.4 mbd, bringing down the amount of oil available in the non-communist world from its October level of 50.8 mbd to 46.4 mbd. The consequences of the embargo threatened the economic growth of Western Europe and Japan. Similarly, the 40 percent jump in oil price caused "gas lines" to emerge in the United States and created economic and social concerns.[17]

16 After Saudi Arabia, Kuwait adopted the second largest cutbacks.

17 Yergin, *The Prize: The Epic Quest for Oil, Money, and Power*, 614–617.

TABLE 4.12 *Saudi Arabia's output change from September 1973 level (1,000 b/d)*

Year	Month	Volume	Change	Percentage Change
1973	September	8,532	–	–
	October	7,796	-736	-8.6
	November	6,267	-2,265	-26.6
	December	6,614	-1,918	-22.5
1974	January	7,513	-1,019	-11.9
	February	7,796	-736	-8.6
	March	8,131	-401	-4.7
	April	8,654	122	1.4
	May	9,014	482	5.7

SOURCE: EIA, MONTHLY ENERGY REVIEW. AVAILABLE, APRIL 2009.

In large part surging oil prices caused the panic, although the oil measures taken by the Arab producers did not directly target oil price but exports. As a consequence of the embargo and cutbacks, shortage and panic in the market affected oil prices: auctions were bidding up the price to an unprecedented level, and the gap between the spot price and the bid price quadrupled.[18] The Saudis, however, were eager to demonstrate that the embargo and cutbacks were enacted to serve a political purpose, not economic gains. The Saudi Oil Minister, Ahmed Zaki Yamani clearly expressed this view. The high prices, in Yamani's words,

> reflect to a large extent the effects of the oil embargo and cutback measures taken by the Arab oil-producing countries, and since these measures are of a political nature, they should not have an economic effect.[19]

The question of the new oil price led to a showdown between Teheran and Riyadh before a compromise was reached at a new spot price of around $12 a barrel.[20]

18 In mid-December at an auction held in Teheran, bid price reached $17.40 a barrel, approximately four times the spot price of $4.75 a barrel.

19 Middle East Economic Survey (December 21, 1973).

20 Whereas the Shah of Iran sought to capitalize on the oil shortage caused by the embargo and cutbacks to raise the price, King Faisal was concerned that high prices could be damaging to world economy, but he also sought to preserve the political objective of the oil

Case Six: The 1978 Camp David Accords

By many measures Saudi Arabia not only emerged intact from the October 1973 War, but also gained financially and politically. Five years later, however, the Saudi ruling family was confronted with another fatal episode in the Arab-Israeli conflict, the Camp David Accords.[21] Although the Saudis were in full support of a peaceful end to the conflict, they favored a peace initiative with a seal of approval from all Arab countries. Syria, the Palestinian Liberation Organization (PLO), Jordan, Iraq, and other Arab countries squarely opposed the Camp David Accords. For its part, Saudi Arabia opposed the Camp David Accords because of domestic and regional security concerns. It was mainly concerned that issues concerning Palestine and the Palestinians were not sufficiently addressed. The Saudis could not support an agreement that did not explicitly recognize the right of the Palestinians for statehood, mandate the full withdrawal of Israeli forces from the occupied territories (the West Bank, Gaza Strip, and Golan Heights), and acknowledge the role of the PLO as a representative of the Palestinian people.

Responses: External Validation

Confronted with the security challenges of the Camp David Accords, the Saudi regime sought to ward off domestic and regional threats. Its relationships with Egypt and the United States were radically modified. During the years before the Camp David Accords, Saudi Arabia acted as the primary financier of Egypt's economic development and military buildup.[22] The disagreement over the Camp David Accords, however, led Saudi Arabia to suspend all economic aid to Egypt. In April, the kingdom agreed to the decision reached by a consortium of Gulf States to dissolve the Gulf Organization for the Development of Egypt (GODE), and in May 1979 it ordered the Arab Organization for Industrialization

---

measures. The disagreement between Teheran and Riyadh over spot price escalated, nearly breaking up OPEC. However, an agreement was finally reached in a historic meeting on 22 December. Yamani was unsuccessful in carrying out instructions of King Faisal to hold the price below $8 a barrel, but the Iranian Oil Minister had also to settle for a price less than $20 a barrel.

21 Camp David Accords refer to a framework of peace in the Middle East and a framework for the conclusion of a peace treaty between Egypt and Israel. Both agreements were signed at the White House by Egypt President Anwar Sadat and Israel Prime Minister Menachem Begin in 1978. For more information on the Camp David Accords, see William B. Quandt, *Camp David: Peace Making and Politics* (Washington D.C.: The Brookings Institution, 1986).

22 According to Jabber, between 1973 and 1976, Egypt received in total about $5,122 billion from Saudi Arabia, Kuwait, the United Arab Emirates, and Qatar. See Paul Jabber, "Oil, Arms and Regional Diplomacy" in Malcolm H. Kerr and El Sayed Yassin, eds., *Rich and Poor States in the Middle East: Egypt and the New Arab Order* (Colorado: Westview Press, Inc.), 428.

TABLE 4.13 *Saudi Arabia oil production (1000 b/d), 1978–1979*

Year	Oil production
1978 September	8,343
1978 October	9,270
1978 November	10,249
1978 December	10,403
1979 January	9,789
1979 February	9,789
1979 March	9,789
1979 April	8,758
1979 May	8,758
1979 June	8,758
1979 July	9,789

SOURCE: EIA, MONTHLY ENERGY REVIEW, APRIL 2009.

(AOI) to liquidate.[23] The Saudis also canceled several arms purchases for Egypt, including the $525 million deal for fifty F-5E jet fighters to be supplied from the United States.[24] In addition to halting economic assistance to Egypt, in the midst of a tight oil market, the kingdom lowered its daily production from 10.4 mbd in December 1978 to 8.5 mbd in April 1979 (see Table 4.13). Although the United States put heavy pressure on Saudi Arabia to raise production, the kingdom refused to cooperate. As a result, the price substantially increased.[25] By the time Saudi Arabia agreed to moderate the price, the Arabian Light oil price had already jumped from $12 a barrel to $18 a barrel.

Both the oil embargo and the Camp David Accords demonstrated the ways in which the Saudi regime used oil resources to bolster its legitimacy at home

23 Saudi Arabia, Kuwait, Qatar, and the United Arab Emirates established the GODE in early 1976, with a capitalization of $2 billion. In 1975 AMIO was established by four Gulf States (Saudi Arabia, Kuwait, Qatar, and the United Arab Emirates) in Egypt with a capitalization of more than $1 billion for the development of military industries.

24 Safran, *Saudi Arabia: The Ceaseless Quest for Security*, 433.

25 William B. Quandt, *Saudi Arabia in the 1980s: Foreign Policy, Security, and Oil* (Washington, DC: The Brookings Institution), 131. Other scholars have argued that the Saudi decision to cut back production was intended to "appease revolutionary Iran." See Safran, *Saudi Arabia: The Ceaseless Quest for Security*, 308. However, what makes the cutback relevant to Camp David is the question of timing. The output ceiling was imposed on March 26, the same day when Egypt and Israel signed the peace treaty. See Holden and Johns, *The House of Saud*, 507.

and in the Arab world. The ruling family deemed external validation necessary measure to deflect possible threats emerging from inside the kingdom and neighboring Arab countries. Contrary to the logic of realism, Saudi Arabia did not only take into account external but also internal security concerns. Placing an oil embargo against the United States was seen by the royal family as a necessary evil in order to demonstrate that the kingdom was in full support of the Palestinian cause. Faisal's decision to enact the oil embargo was at odds with the expected security behavior. As pointed out by Holden and Johns,

> Kissinger had difficulty in understanding why, out of self-interest, Feisal could not or would not align more fully with the U.S. and thus fall with more tidiness into his conceptual framework of a worldwide balance of power between the two super-powers.[26]

Similarly in the case of the Camp David Accords, without a just and fair solution to the Palestinians and the "liberation" of Jerusalem, public opinion at home and throughout the Muslim world would swing against the royal family. Therefore, the royal family acted on the idea that Saudi Arabia security interests were best served by expressing its support to Arab consensus, even at the cost of its friendship with the United States and relationship with Egypt.

Case Seven: The Iranian Revolution

In early 1978, Iran, the only Shia majority country in the Muslim world, plunged into a social and political upheaval. The explosion of the Iranian Revolution brought down the pro-American regime of the Shah. The arrival of the Iranian Revolution in 1978 drastically altered the Saudis' sense of security. The propaganda attack on the Saudi royal family's Islamic credential was most concerning. As Long remarks, "This is extremely disconcerting to the Saudi regime, which since the Wahhabi revival nearly 250 years ago has considered itself to be the right wing in terms of Islamic puritanism."[27]

Although the propaganda attacks transmitted by Radio Teheran and Radio Ahwaz aimed to incite rebellion in the Shia community, they also intended to

26 Holden and Johns, *The House of Saud*, 348. Other authors have pointed out that King Faisal came under heavy pressure from inside and outside the kingdom to unleash the oil weapon. See Yergin, *The Prize: The Epic Quest for Oil, Money, and Power*, 595.

27 David E. Long, "The Impact of the Iranian Revolution on the Arabian Peninsula and the Gulf States," in John L. Esposito, ed., *the Iranian Revolution: Its Global Impact* (Florida: Florida International University Press, 1990), 108.

rouse opposition among the Sunni population. Corruption and embezzlement of public wealth, grievances over modernization, and the perceived pro-Western oil policy angered many extremists.[28] This public discontent was fanned by Iranian propaganda, which did not cease taking the opportunity to undermine the reputation of the royal family for wasting public money on "gambling, drinking parties and shameless dancing and orgies."[29] The Iranian propaganda blasted the royal family for the un-Islamic nature of their rule, amoral life style, complicity with the infidels and Zionism, and worthlessness as the guardian of Islamic holy places.

The pilgrimage to Mecca (*hajj*) was employed as another means to disrupt order and embarrass the Saudi ruling family.[30] Khomeini instructed his followers to consider *hajj* not simply as a religious obligation, but also to give equal weight to its social and political aspects.[31] Following the instruction of Khomeini, in 1981 Iranian pilgrims clashed with Saudi security forces. As a result almost 10,000 to 12,000 pilgrims were arrested and deported. Similarly in 1982, "one hundred thousand Iranian pilgrims shouting revolutionary slogans under placards carrying Khomeini's picture clashed with Saudi police."[32] In 1986 the Saudi government sentenced Iranian pilgrims to prison for attempting to smuggle explosives and weapons into Saudi Arabia. In 1987, the clash between the Saudi forces and thousands of Iranian pilgrims left 402 people dead.

---

28 For instance, Said al-Nasir, the leader of the Union of the People of the Arabian which took responsibility for the attack on the Grand Mosque, charged the regime with "depleting the wealth of future generations." Middle East Contemporary Survey, Volume Four (1979–80), 691. It is noteworthy that Saudi officials have acknowledged that their oil policy was "beneficial" to America, and "that they did not have to produce as much oil as they did, or to be moderate in pricing it." See Middle East Contemporary Survey, Volume Four, (1979–80), 54. Similarly, as pointed out by Quandt, there is a strong economic case to be made for producing less oil, thus prolonging the life of oil resources for the benefits of future generations. The main point of this argument is that increases in oil prices can accommodate the budget needs without producing more than 5 mbd to 6 mbd. See William B. Quandt, *Saudi Arabia in the 1980s: Foreign Policy, Security, and Oil*, 123.

29 Middle East Contemporary Survey, volume four (1979–80), 694.

30 The *hajj* is a sacred mission to Mecca obliged to be performed by every able, adult Muslim as one of the five pillars of Islam.

31 Middle East Contemporary Survey, volume six (1981–82), 286.

32 Sandra Mackey, *The Iranians: Persian, Islam and the Soul of a Nation* (New York: Penguin Group, 1996), 312.

Rebellion in the Shiite Eastern Province

In the course of the traditional procession of the *Ashura*, thousands of Shia clashed with security forces on 28 November 1979. The demonstrators shouted anti-royal family slogans, demanded that the regime stop supplying oil to the United States, end Sunni-majority discrimination against the Shi`a community, and support the Iranian Revolution. The riot was dissolved with the help of 20,000 security forces.[33] But a few months later, in early February 1980, the Shia community of Hasa staged a second riot. Once again the demonstrations took a violent course, with the burning of cars and buses and official buildings. The clash with security forces left four people dead. The disturbances in Hasa were particularly concerning, because the region is home to oil facilities. Thus, there were concerns that oil fields may be targeted for sabotage activities, especially if Iranian radicalism took hold among the Shia community. Iran was actively encouraging a rebellion against the Saudi royal family, with the circulation of anti-royal family leaflets and cassette recordings of Khomeini's religious speeches.[34]

Responses: Internal Balancing and Internal Validation

To counter internal threats posed by the Iranian regime, the kingdom used internal balancing as well as internal validation. Measures of internal security were increased with the establishment of a special anti-terrorist unit armed with modern equipment, including helicopters, armored personnel carriers and sophisticated electronic systems. As shown in Table 4.10, the budget allocation to internal security was largely increased in the years after the Iranian revolution. For instance, allocations to the Ministry of Interior increased from 5.8 trillion SR (1976–77), to 9 trillion SR (1978–79), and then to 20 trillion SR (1984–85). Other means of increasing security included the large-scale deportation of illegal immigrants, to the extent that in February and March 1980 about 20,000 aliens were deported. In addition to these security measures, the royal family made efforts to prop up its image at home. Seeking to alleviate concern and accommodate interest of the Shi`a community, various Saudi princess met community leaders and promised to improve the socio-economic conditions of the Eastern Province. In the fiscal year 1980–81 the government more than doubled the budget allocation to the Eastern Province, from 78 million SR to 163 million SR in the fiscal year 1979–80.[35]

---

33 Middle East Contemporary Survey, volume four (1979–80), 688.

34 Middle East Contemporary Survey, 4 (1979–80), 689.

35 Ministry of Economy and Planning, *Annual Statistical Books, 1975 and 1980*.

Case Eight: The Iran-Iraq War, 1980–1988

The outbreak of the Iran-Iraq War in 1980 discharged several security concerns to the stability of the Saudi regime and the security of the kingdom.[36] First, there were worries about the security of oil installations, and the blockade of oil shipments in the Strait of Hormuz. The second threat was the spread of Iranian influence in the Arabian Peninsula. The refusal of Iran to accept repeated calls for ceasefire and its launch of massive attacks into Iraqi territory confirmed Saudi suspicion of Khomeini. The Saudis feared that Khomeini's intension was to spread Shia influence in neighboring countries, not only by means of propaganda but also by force. Khomeini was blunt about his intentions. In his words,

> Every day of the war we had blessing, which we utilized in all aspects. We exported our revolution to the world through the war…it was through the war that we recognized our enemies and friends.[37]

Third, not only did the war burden Saudi Arabia financially but also increase the kingdom's perception of insecurity. A victorious Iran would strengthen the impulse of the Iranian Revolution, enhance its prestige among the Shia and Sunni communities in the Arab countries, and improve its position vs. Saudi Arabia in the bid for leadership in the Muslim world. On other hand, a victorious Iraq would improve the position of Saddam's secular regime in the Gulf. But the secular characteristics of the Iraqi regime did not pose the same acute threat as the Islamic-colored Iranian threat, making Iraq the lesser of two evils. Hence, the Saudi regime, while officially claiming neutrality, adopted in practice a pro-Iraq policy.

Responses: Internal and External Balancing

To confront the multiple threats of the Iran-Iraq war, Saudi Arabia resorted to internal balancing, and external balancing with Iraq and consuming countries. In the case of internal balancing, the Saudi regime sought to enforce its defense capabilities. It turned to the United States and Britain for arms. Shortly after the outbreak of the war, the United States deployed to the kingdom an air defense aircraft system (WACS) that could detect approaching warplanes from

---

36 The main reason for the war was disagreement between Iran and Iraq regarding the division of the waters of the Shatt al-Arab River.

37 Quoted in Rouhallah K. Ramazani, "Iran's Export of the Revolution: Politics, Ends, and Means," in John L. Esposito, ed., *the Iranian Revolution: Its Global Impact* (Florida: Florida International University Press, 1990), 52.

250 miles in all directions.[38] Reflecting the Saudi royal family's concerns over Iranian strikes against oil facilities, the kingdom concluded with Britain an arms package deal worth $4.5 billion, which included 72 Tornado combat jets and 60 Hawk and PC9 trainers.[39] In general, spending on defense and security during the Iran-Iraq swelled, as shown in Table 4.10.

As Western powers were ready to guard the kingdom against Iranian attacks, the Saudi regime strove to ensure stability in the oil market. The Iran-Iraq War was a major threat to the stability of the oil market. Since both countries were major oil producers and relied heavily on oil revenue to support their war efforts (see Table 4.14), oil installations were singled out as strategic military targets.[40] During the course of the war, both countries experienced drastic declines in oil production. For Iran, the average annual production bottomed to 29 percent of its level of 1978. Likewise, the Iraqi average annual production declined to 39 percent of its level of 1978. In order to make up for oil shortages Saudi Arabia raised its production level to near the maximum, producing over 9 mbd in 1980 and 1981 (see Table 4.14).

Finally, the Saudi regime sought to influence the outcome of the war by providing Iraq with financial support and exerting financial pressure on Iran. Saudi Arabia's support for Iraq was largely facilitated by oil money as well as oil production and pricing strategies. In terms of financial aid, Saudi support to Iraq, in the form of petrodollars, amounted to more than $25 billion.[41] To the disadvantage of Iran, Saudi Arabia advocated an oil policy of higher output

---

38 In addition to supplying AWACS planes, the United States offered to set up a joint naval force. The Saudis, however, turned down the offer for fear that it might inflame the Saudi-Iran relationship.

39 Karen De Young, "British Get $4.5 Billion Plane Order; Saudi Purchase Had Green Light From White House," *Washington Post*, September 27, 1985.

40 As Cordeman and Wagner points out, "Iran and Iraq's relative ability to mobilize their economy and oil wealth was a major factor shaping the entire course of the war." See Anthony H. Cordesman and Abraham R. Wagner, *The Lessons of Modern War: The Iran-Iraq War*, Volume II (Colorado: Westview Press, 1990), 45. To hinder Iran's oil exports, Iraq launched repeated attacks on the Iranian main oil port of Kharg, triggering what is called the "Tanker War." Through the Kharg terminal Iran exported 1.5 mbd to 2 mbd. Iran responded, in like manner, bombing Iraq's main port of Faw. As oil production figures show in table 4.14, during the first three years of the war, Iranian attacks proved to be successful in curtailing Iraqi oil production. For more information on the oil war, see Cordeman and Wagner, *The Lessons of Modern War: The Iran-Iraq War*, 90–92.

41 "$5,843,287,671 as nonreturable aid; $9,246,575,342 as easy payment monetary loans; $95,890,410 as development loans; $3,739,184,077 for military and transport equipment; $6,751,159,583 as oil aid; $16,772,800 as the value of industrial products for reconstruction of Basra; $20,266,667 for SABIC entitlements to Iraq, and $21, 333,333 for 270 tractor trucks

TABLE 4.14 *Crude oil exports (1,000 b/d)*

Year	Iran	Iraq	Saudi Arabia
1978	4,447	2,384	7,706
1979	2,407	3,247	8,818
1980	797	2,482	9,223
1981	715	872	9,018
1982	1,623	846	5,639
1983	1,719	702	3,921
1984	1,522	867	3,187
1985	1,568	1,085	2,151
1986	1,454	1,394	3,266
1987	1,710	1,717	2,417
1988	1,696	2,095	3,030

SOURCE: OPEC ANNUAL STATISTICAL BULLETIN, 2005.

and lower price. Such a policy benefited Iraq because it had secured access to exporting oil to the world. The newly expanded oil pipeline running through Turkey increased Iraqi oil output from 1 mbd to 1.5 mbd. With the extra 0.5 mbd Iraq's oil exports were about 1.2 mbd above its quota of 1,466,000 bd.[42] In addition to the existing pipeline that moved 500,000 b/d of Iraqi oil (IPSA 1), Saudi Arabia also agreed to a second pipeline (IPSA 2) in which an additional 1.5 mbd of Iraqi oil could be moved to the Saudi port of Yanbu on the Red Sea. Additionally 200,000 barrels of crude and refined oil were trucked through Turkey and Jordan.[43] On the other hand, Iran's ability to produce and sell more oil to the outside world was handicapped by Iraq's heavy bombardments of Iranian oil installations and lack of secure outlets.

The major shift in Saudi Arabia's oil policy occurred in September 1985, when oil exports were priced on the basis of "netback," instead of spot price. By charging consumers the gross value of refined products minus refining costs and rate of transportation from producing country to refinery, Saudi Arabia set

for spreading asphalt." See Simon W. Murden, *Emergent Regional Powers and International Relations in the Gulf: 1988–1991* (Reading: Ithaca Press, 1995).

42 John Hooper, "Don't rock a boat making headway: The latest Opec meeting," *Guardian*, June 26, 1987.

43 Youssef Ibrahim, "New Pipelines Are Reducing Persian Gulf's Strategic Role," *New York Times*, October 07, 1987.

the stage for a price collapse. On this basis of pricing strategy, in September the kingdom signed agreements covering 1 mbd.[44] The enacting of the netback took place around the same time as the Iran-Iraq War had decisively turned against Iraq. In the second half of 1985, Iran prepared for a major offense to capture Faw Peninsula and Umm Qasr.[45] This massive mobilization convinced the Saudis that Iran aimed to break through the Iraqi defense, so long as the Iranian leadership had access to a large amount of oil revenue.[46] About the same time, Iraq and Iran stepped up their efforts to halt each other's oil output. Iraq intensified air raids against Iran's main offshore oil terminal on Kharg Island. For its part, Iran threatened to halt all Arab oil exports passing through the Hormuz Strait, if its oil production was significantly reduced.[47] By February 1986, the kingdom had signed contract agreements of supplies covering 3.5 mbd priced on netback basis. Subsequently, in March, oil prices plunged to 13 dollars per barrel, and then to $9 a barrel in July (see Table 4.15), and were predicted to even reach $5 a barrel.[48] The effect of Saudi Arabia's oil policy on the Iranian war chest proved to be more powerful than Saddam's bombardments of the Kharg terminal.[49] Iran suffered tremendously from the Saudi

44 Parra, *Oil Politics: A Modern History of Petroleum*, 284–287.

45 Faw was significant as the main port for oil export. It was destroyed in the early years of the war, and later was occupied by Iran. The port city of Umm Qasr, located only 16 kilometers from Faw and home to an Iraqi naval base, was able to hold out against the Iranian attacks.

46 The massive mobilization efforts included recruiting women as rear-area military personnel, building up new road infrastructure for military usage, turning civil servants into combatants, calling up new volunteers, and buying chemical defense equipment. See Cordesman and Wagner, *the Lessons of Modern War: The Iran-Iraq War*, 217–224.

47 David B. Ottaway, "Persian Gulf War Posing New Threat To U.S. Interests; Iraq Targets Iran's Main Oil Terminal," *Washington Post*, September 27, 1985.

48 The fallout of the dramatic slide in the price led to a turbulent year for OPEC members, as the crisis split the Arab group into two sides. On one side were the four Arab Gulf States led by Saudi Arabia which favored higher output and lower oil price. On the opposite side were Libya and Algeria sided with Iran and sought production cuts designated to bolster oil price. See John Tagliabue, "An OPEC Meeting Seems At Impasse," *New York Times*, August 03, 1986.

49 There may have been an economic motive behind the new Saudi oil policy, but as Hunter points out: "Whether or not the desire to hurt Iran economically was the principal motive behind the Saudis' oil policy, they could not miss its devastating impact on Iran, coming at the time of the Iranian capture of Faw and growing Saudi anxiety about Iraq's fate." See Shireen T. Hunter, *Iran and the World: Continuity in a Revolutionary Decade* (Bloomington: Indian University Press, 1990), 119. The Iranian Deputy Foreign Minister at the time, Ali Mohammed Besharati, accused Saudi Arabia as "guilty of the greatest treason ever committed against the oppressed and deprived countries." See Ibid., 120. Moreover, the

TABLE 4.15 *Spot OPEC reference basket price and its components (US dollars per barrel), 1985–1986*

Month-year	Iran heavy	Iraq basrah light	Arabian light	OPEC reference basket
Jan-85	25.9	27.6	27.9	27.1
Dec-85	25.4	26.6	27.9	26.5
Jan-86	23.5	24.0	23.8	22.2
Feb-86	17.6	18.9	15.9	16.0
Mar-86	12.7	13.1	13.1	12.8
Apr-86	9.7	11.5	11.3	11.6
May-86	11.4	12.4	12.6	12.9
Jun-86	10.0	10.7	10.8	11.2
Jul-86	7.6	8.3	8.5	9.0

SOURCE: OPEC STATISTICAL BULLETIN, VARIOUS YEARS.

downward pressure on the oil price, seeing its revenues declining by more than half, from $13 billion in 1985 to $5,900 billion in 1986 (see Table 4.16). Ultimately it was the Saudi oil policy that drained the war chest of Iran.

Conclusion

In the early 1970s, the Saudi state took full control over its oil resources from ARMACO. Given that the kingdom possessed a vast amount of proven reserves and considerable production capacity, the Saudi state acquired significant influence in the oil market. With respect to consumers, its influence was magnified by the strong world demand for oil, which lasted until the early 1980s. After 1981 demand for oil steadily declined, especially in OECD countries. Many oil producers, however, did not curtail their output. With weak demand and excess supply, oil rentier states were particularly vulnerable to financial crisis. Therefore, in this period, both consumers and producers were asymmetrically dependent on Saudi Arabia.

---

crisis within OPEC in the year 1986 clearly showed that Iran (as well as Algeria) believed the purpose of the Saudi oil policy was to deprive Iran of oil revenues in order to cripple its war efforts.

TABLE 4.16 *Values of petroleum exports (Million US dollars)*

Country	1980	1981	1982	1983	1984	1985	1986	1987	1988	1989	1990
Iran	11,693	10,047	18,690	20,273	15,713	13,012	5,900	9,400	8,419	11,315	16,831
Iraq	26,096	10,039	9,933	7,816	8,863	10,097	6,905	9,416	9,312	11,876	9,594
Saudi Arabia	1,01,421	1,11,546	73,263	44,829	36,282	25,930	18,060	20,427	20,205	24,095	40,129
TOTAL OPEC	2,79,056	2,54,920	2,01,716	1,61,605	1,51,318	1,32,567	78,134	94,296	88,415	1,10,528	1,51,020

SOURCE: OPEC ANNUAL STATISTICAL BULLETIN, 2005.

The Saudi regime engaged in internal validation and internal balancing at a higher level than in the 1950s and 1960s. As far as internal validation is concerned, the state initiated a series of five-year development plans, with the following implications. First, a significant number of the working force became dependent on state employment, either in the civil administration or in the public sector. Second, direct distribution of wealth improved people's living standard and income through grants, subsidies, and welfare programs. Third, indirect distribution of oil wealth through massive infrastructure development became the primary catalyst behind economic growth of the private sector. In terms of internal balancing, huge amounts of the revenues were allocated to the development and support of military and internal security apparatuses.

The responses to major threats in this period demonstrate the usage of oil resources as a mechanism for external validation and external balancing. The oil embargo of 1973 and the response to the Camp David Accords demonstrated the deployment of external validation in order to maintain societal support and avoid domestic instability. In the case of the 1973 oil embargo, Saudi Arabia, in concert with Arab oil-producing countries, cut oil production by 25 percent and enacted a ban on all oil shipments to the United States, the Netherlands, South Africa, Rhodesia, and Portugal. Likewise, to avoid isolation in the Arab world, the Saudis responded to Camp David by imposing a ceiling on supply in the midst of tight oil market turmoil.

To confront the threat of Revolutionary Iran, Saudi Arabia balanced with Iraq and the Western consuming countries, especially the United States. While the United States provided military assistance and security guarantees, Saudi Arabia overproduced to make up for oil shortages caused by the missing Iraqi and Iranian oil. At the same time, it supported Iraq with financial aid and oil supplies. Moreover, the kingdom turned oil price into an economic weapon against Iran. In order to deprive the Iranian regime from oil revenues, the kingdom resorted to producing more oil at lower prices.

CHAPTER 5

The Bust (1985–2000)

Unlike the previous decade, the second half of the 1980s and most of the 1990s were marked by financial constraints. With oil prices at a nadir, the kingdom endured a perennial deficit, lasting until the end of the century. The era of oil bust strained the regime, but it did not alter the mechanisms of security and stability. In this chapter I analyze how the bust led to a new pattern of dependence, which in turn shaped internal and external security strategies. Similar to the previous two chapters, this chapter is divided into four three. The first section covers measures of dependence, global supply and demand of oil as well as Saudi Arabia's exports and proven reserves. Section two examines internal security strategies: internal validation and internal balancing. Section three describes security strategies employed in responses to specific threats.

Pattern and Domain of Dependence in the Bust Era (1985–2000)

Oil Market: Supply and Demand

The 1980s arrived with soft market conditions, exerting heavy downward pressure on oil prices. Despite efforts to hold up prices, in 1986 the price collapsed, and it remained low until the end of the century, except for certain occasions in which prices temporarily spiked. Although since 1985 oil consumption continued to increase (see table 5.1), several reasons led to a soft oil market during the second half of the 1980s and most of the 1990s: a significant factor was the slowing down of the global economy, which caused demand for oil to decline. Other factors were a byproduct of the oil shocks of the 1970s. The high prices of the 1970s dealt a blow to OECD economies. A decline in economic activities in OECD countries in turn translated to less oil consumption. Moreover, in many OECD countries the oil shocks led to enacting of policies that aimed to reduce reliance on imported oil by means of increasing efficiency, stimulating alternative sources of energy, and expanding local exploration and production.[1] The 1990s arrived with a stronger demand for oil, but did not move far above the 1979 level. Demand accelerated in the second half of the 1990s. But the growth in demand did not shift the balance of power in favor of producers. As Feld

1 Siamack Shojai and Bernard S. Katz, ed., *The Oil Market in the 1980's: A Decline of Decline* (New York: Praeger Publishers, 1992), 32–33.

 | DOI 10.1163/9789004277731_007

TABLE 5.1 *World oil consumption (Million b/d), 1986–2000*

Year	Total OECD	Total Non-OECD	World
1986	39	23	62
1987	39	24	63
1988	41	24	65
1990	42	25	67
1992	43	25	67
1994	44	24	69
1996	46	26	72
1998	47	27	74
2000	48	29	77

SOURCE: EIA, ANNUAL ENERGY REVIEW 2007.

points out, with the development of oil spot and futures markets, the price was set by open markets as much as by oil producers. At the same time, oil-consuming countries had developed massive strategic oil reserves.[2]

Despite the modest recovery in demand, the market remained weak. Over the last 15 years of the twentieth century the price of oil remained under $20 per barrel, except during the months of the Iraqi invasion and occupation of Kuwait, when oil prices hovered around $23 a barrel. The serenity of the market can be explained partly by weak demand, especially during the first half of the 1990s. Equally important was the expansion in exploration and development, which had been instigated by high prices of 1970s and early 1980s. Hence, world proven oil reserves grew from 643 billion barrels in 1980 to 1985 billion barrels in 1990 (53 percent), and then to 1 trillion barrels in 1995. This gigantic accumulation in reserves took place in spite of the 331 billion barrels that the world had produced over the same period. Actual production did not go over the 1980 level of 21 billion barrels until the second half of the 1990s, when it reached 24 billion barrels. In other words, there was abundance of supplies. Table 5.2 shows changes in oil production by region over this period. Although production did not rise, there were slight increases across the region. Most notable is the progressive growth in major consuming markets, such as Western Europe (60 percent) as well as Asia and the

2 Lowell S. Feld, "Oil Markets in Crisis: Major Oil Supply Disruptions Since 1973," in Siamack Shojai, ed., *The New Global Oil Market: Understanding Energy Issues in The World Economy* (WestPort: Praeger Publishers, 1995), 110–111.

TABLE 5.2 *Oil production by region (1000 b/d), 1986–2000*

Year	North America	Latin America	Eastern Europe	Western Europe	Saudi Arabia	Africa	Asia and Pacific	Middle East[1]
1986	9,810	6,059	12,309	3,931	4,784	5,154	5,750	7,319
1988	9,420	6,114	12,368	4,046	5,100	4,961	5,854	9,050
1990	8,562	6,872	11,275	4,085	6,413	5,967	6,270	9,664
1992	8,407	7,271	8,843	4,556	8,332	6,329	6,358	9,231
1994	8,012	7,556	7,167	5,568	8,049	6,147	6,630	10,760
1996	7,866	8,148	6,938	6,160	8,102	6,426	7,022	10,910
1998	7,679	9,467	7,088	6,079	8,280	6,674	7,044	12,835
2000	7,213	9,317	7,631	6,288	8,095	6,746	7,253	13,316

(1) Saudi Arabia is not included.
SOURCE: OPEC, ANNUAL STATISTICAL BULLETIN, 2007.

Pacific (26 percent). The largest growth, however, came from the Middle East, including Saudi Arabia (77 percent). Therefore, unlike the tight market of the 1970s, in the last fifteen years of the twentieth century the market was driven by supply, not by demand.

Saudi Arabia: Oil Reserves and Exports

Despite its sizable production over several decades, Saudi Arabia's proven reserves and export volumes did not decrease in the last fifteen years of the century. Its proven reserves in the 1985–2000 were comparable to the levels of the 1950s, and 1960, and 1970s. Despite producing 133 trillion barrels of oil in the period of 1960–1985, the kingdom's proven reserves have increased from 53 billion barrels in 1960, to 140 billion barrels in 1970 (167 percent), and then to 171 billion barrels in 1985 (21 percent).[3] As shown in table 5.3, relative to other regions, the kingdom possessed the lion's share of proven reserves.

Given its considerable size of reserves, between 1981 and 1985 the kingdom adopted a swing producer strategy. As a swing producer, the kingdom did not have a quota. In times of strong demand the swing producer raises production as much as needed. However, as the market de-accelerates the swing producer significantly reduces production. Having acted as a swing producer during the period 1980–1985 the kingdom lowered its production by 68 percent. The massive cutbacks caused the kingdom to lose substantial market share, not to mention oil revenues. By 1985 its share of the world export had shrunk, from 30

3 OPEC, Annual Statistical Bulletin, 2007.

TABLE 5.3 *Proven oil reserves by region as percentage of total world reserves, 1985–2000*

Region/Country	1985	1990	1995	2000
North America	4.5	3.2	2.7	2.5
Latin America	15.7	12.5	13.1	11.2
Eastern Europe	8.5	6.0	5.8	8.7
Western Europe	1.9	1.7	2.1	1.7
Saudi Arabia	22.7	26.4	25.9	24.1
Africa	7.4	5.9	7.0	8.6
Asia and Pacific	4.9	3.5	3.5	3.6
Middle East-SA	34.4	40.8	40.0	39.6

SOURCE: OPEC, ANNUAL STATISTICAL BULLETIN, 2007.

percent in 1980 to 10 percent in 1985, and revenues fell from $102 billion $27 billion over the same period. As a result, the economic and political situations were no longer sustainable (as will be shown in the next section), and the kingdom decided it was time to adopt a new oil policy. In 1985 the kingdom abandoned the role of a swing producer and instead enacted a policy of pricing its oil on a "netback" basis. The shift in policies aimed to recover market share by increasing export levels (see table 5.4). After its nadir of 2.5 mbd in 1985, output rebounded to 3 mbd in 1986, then to 4.5 mbd in 1990, and then to 6.5 mbd in 1992. Between 1990 and 2000 the kingdom's export volume as percentage of total OPEC exports averaged yearly 28 percent. The strategy of netback helped to recover market share, increasing the kingdom's market share from 10 percent in 1985 to 16 percent of world export in 2000.

Patterns of Dependence during 1986–2000

The economic recovery in the 1990s strengthened demand for oil, but overall demand did not outpace supply. The 1990s was a consumer market with a large excess capacity in the world. During this period the influence of the kingdom increased with respect to other rentier states. Because oil prices were low and Saudi unparalleled production capacity, Saudi Arabia had the capability to pressure prices either upward or downward. Already in the range of $13 to $24 less than the price levels of the 1975–1985 period, further drops in prices were bound to generate major fiscal imbalance for oil rentier states. The kingdom, therefore, held some leverage over other rentier states. Under these conditions the kingdom was asymmetrically dependent on consuming countries but producers were dependent on Saudi Arabia.

TABLE 5.4 *Oil exports by region (1000 b/d), 1986–2000*

Year	North America	Latin America	Eastern Europe	Western Europe	Saudi Arabia	Africa	Asia and Pacific	Middle East[1]
1986	736	2,818	2,535	2,467	3,266	3,766	2,037	5,618
1988	864	2,910	2,823	2,451	3,030	3,486	1,918	6,880
1990	763	3,149	2,133	2,565	4,500	4,348	2,099	7,579
1992	926	3,498	1,480	3,073	6,582	4,620	2,229	6,805
1994	1,081	3,872	1,878	4,029	6,234	4,422	2,292	7,626
1996	1,182	4,626	2,069	4,524	6,109	4,892	2,204	7,721
1998	1,259	5,254	3,045	4,527	6,390	5,274	2,193	9,212
2000	1,231	4,906	4,145	4,991	6,253	5,209	2,230	9,835

(1) Saudi Arabia is not included.
SOURCE: OPEC, ANNUAL STATISTICAL BULLETIN, 2007.

Before moving to a discussion on security strategies in the next section, in what follows I briefly discuss the financial situation of the kingdom during this period. As a result of weaker oil prices and slower demand, the financial position of the Saudi state was weaker in comparison with the 1970s and early 1980s. As shown in table 5.5, as a result of the of the Iraq invasion of Kuwait, oil prices briefly jumped to $20 a barrel. However, it rapidly dropped to $15 a barrel in 1993, and five years later, in 1998, it plunged to $12 a barrel. As oil prices declined, revenues also declined. In contrast to an average of 220 billion SR during the period 1980–1985, oil revenues plummeted to an average of 64 billion SR in the second half of the 1980s. To boost revenues, the kingdom abandoned its role as a swing producer and adopted an oil policy aimed at securing larger market share, instead of defending a certain oil price. It aggressively increased production. The primary objective of this policy was to compensate for loss of revenues caused by depressed oil prices. In contrast to the first half of the 1980s, when Saudi Arabia cut back production in order to moderate prices, in the second half of the 1980s and throughout the 1990s market share was a top priority, not price. Therefore, the kingdom did not adjust production to prop up the slumped price in 1993 and 1998. Instead, oil exports were raised to more than 6 mbd throughout the 1990s.

Because oil revenues provided the overwhelming part of total revenues, it was inevitable that the budget would be affected as the prices decreased. As shown in table 5.6, the budget turned to a deficit – and even exploded during the Gulf War. The fiscal year 1990, which was merged with 1991, reflects a jump in the deficit as a consequence of extra military spending. Initially

TABLE 5.5 *Saudi Arabia oil revenue, 1986–2000*

Year	Oil revenue (Million SR)	Oil revenue as perventage of total revenue	Oil price Arabian Light	Oil export (1,000 b/d)
1986	42,464	56	13.73	3,266
1987	67,405	65	17.23	2,417
1988	48,400	57	13.40	3,030
1989	75,900	66	16.21	3,336
1990/1991*	2,46,297	78	20.82	4,500
1992	1,28,790	76	17.94	6,582
1993	1,05,976	75	15.68	6,293
1994	95,505	74	15.39	6,234
1995	1,05,728	72	16.73	6,291
1996	1,35,982	76	19.91	6,109
1997	1,59,985	78	18.71	6,185
1998	79,998	56	12.20	6,390
1999	1,04,447	71	17.45	5,720
2000	2,14,424	83	26.81	6,253

* Budget allocation for fiscal year 1991 was amalgamated with the budget for 1990.
SOURCE: SAUDI ARABIAN MONETARY AGENCY (SAMA), TABLES OF ANNUAL REPORT; FOR OIL EXPORTS: OPEC, ANNUAL STATISTICAL BULLETIN, 2007.

the government responded to the deficit by lowering public spending. Thus, spending was reduced from 284 billion SR in 1981 to 2137 billion SR in 1986. After 1986, however, the policy of reducing public spending was reversed. Only in certain years, slight reductions were made when prices plunged; otherwise, in spite of the running deficits, government expenditures fluctuated, but did not sink to the 1986 level (see table 5.6). Meanwhile, the budget swelled.

To balance the budget, the government needed to increase revenues and reduce expenditures. Both measures, however, were deemed to be politically risky. Drastic reductions in expenditures were perceived to undermine the tacit pact between the state and society, in which economic rewards were exchanged for acquiescence and support for the regime. As for revenues, the kingdom could do very little. With oil production at more than 6 mbd, the kingdom could not risk higher exports, lest the oil price would collapse. Instead, to help finance the deficit, the government turned to foreign reserves as well as domestic and

TABLE 5.6 *Budgetary deficit and surplus (Million SR), 1986–2000*

Year	Total revenues	Total expenditure	Deficit/surplus
1986	76,498	1,37,422	-60,924
1987	1,03,811	1,84,919	-81,108
1988	84,600	1,40,856	-56,256
1989	1,14,600	1,54,870	-40,270
1990/1991	3,16,639	4,87,425	-1,70,786
1992	1,69,647	2,38,987	-69,340
1993	1,41,445	1,87,890	-46,445
1994	1,28,991	1,63,776	-34,785
1995	1,46,500	1,73,943	-27,443
1996	1,79,085	1,98,117	-19,032
1997	2,05,500	2,21,272	-15,772
1998	1,41,608	1,90,060	-48,452
1999	1,47,454	1,83,841	-36,387
2000	2,58,065	2,35,322	22,743

SOURCE: SAUDI ARABIAN MONETARY AGENCY (SAMA), ANNUAL REPORTS, 1986–2000.

foreign debt. By 1994, domestic government debt reached $100 billion, approximately 77 percent of the country's GDP, and that proportion was projected to reach 110 percent by 2000.[4] Although the government was loath to borrow from international banks, in 1997 J.P. Morgan, a premier investment bank, arranged a syndicated loan of $4.3 billion.[5] In 1998, when the oil price plummeted to as low as $12 a barrel, the government turned to Abu Dhabi for a loan in the amount of $5 billion. Most significantly, however, as shown in the next sub-section, the government did not institute taxes nor did it drastically cut expenditures on social services. Some subsidies were cut. For example, agriculture subsidies were reduced; consumer products subsidies, such as gasoline, electricity, water, were reduced; and government services fees were increased. But as Gause puts it: "These baby steps alone cannot put the Saudi fiscal house in order."[6]

4 Elaine Sciolino, Jeff Gerth and Douglas Jehl, "Saudi Kingdom Shows Cracks U.S. Aides Fear," *New York Times*, June 30, 1996.

5 James R. Kraus, "U.S. Banks See Borrowing Rise In Persian Gulf," *The American Banker*, March 5, 1998.

6 F. Gregory Gause III, "Saudi Arabia Over a Barrel," *Foreign Affairs*, no. 79 (May/June 2000): 84–94.

Internal Validation and Internal Balancing in the Bust Era

Internal Validation

In the light of the drastic reduction of oil revenues, the government was forced to make some budgetary adjustments and cutbacks. To understand the political reasons that certain cutbacks were made but not others, it is inadequate to only examine the total government expenditure in years of slumped oil prices. In actuality, cutbacks were selectively made in order to minimize the negative economic effect on the population, especially in the area of employment, welfare services, and subsidies.[7] To appreciate the role of expenditures as a mechanism of mitigating domestic discontent, it is necessary to distinguish between and separately examine the two branches of government spending, current and capital expenditure. Current expenditure refers to recurring expenditures on goods and services consumed during a short period, often the current year, including defense, wages and compensations, subsidies, and welfare programs. On the other hand, capital expenditure refers to items that have a longer lifetime, for instance, building roads, schools, hospitals, and purchasing new equipment. In terms of usefulness to the economy, capital expenditures provide a long-term value, increasing the productivity of the economy. In contrast, current expenditure has no lasting effect on the productivity of the economy. In other words, capital expenditure is investment in the future. In light of this distinction, the trends of current and capital expenditures during the last three decades are shown in figure 5.1. As the figure 5.1 shows, both current and capital expenditure rose, but the situation changed in the time of bust when oil prices declined. After 1986 current expenditure continued to rise, but capital expenditures rapidly declined.

There are two plausible explanations for the dissimilarity in the trend of current and capital expenditures. One possible explanation is that many of the infrastructure projects were completed, and therefore it was feasible to reduce the amount allocated to project expenditures. Though the government had planned and allocated project expenditures (see table 5.7), it was forced to revise the planned level of distribution because of the financial constraints. A more probable explanation is that the government gave higher priority to current expenditure, because it has more immediate impact on issues of domestic stability. This reasoning is more credible if one recognizes the two

7 Some analysts have erroneously concluded that there is a direct link between deficit and political turmoil, or total expenditures and turmoil. Fandy for example makes the link between the deficit and turmoil. But as we have seen, the Saudi government continued to spend, despite the deficit. See Fandy, *Saudi Arabia and The Politics of Dissent*, 26–29.

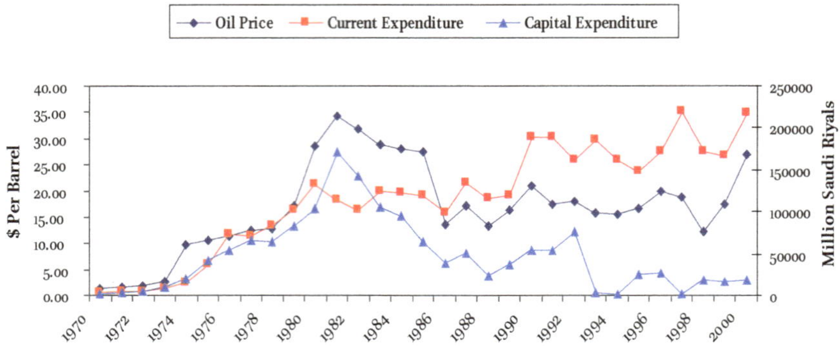

FIGURE 5.1 *Oil price vs. Current and capital expenditures*
SOURCE: CURRENT AND CAPITAL EXPENDITURES: SAUDI ARABIAN MONETARY AGENCY (SAMA); OIL PRICE: BRITISH PETROLEUM, STATISTICAL REVIEW OF WORLD ENERGY JUNE 2008.

distinct implications of wealth distribution, which are closely connected to the type of expenditure: to co-opt influential groups, and to buy political passivity of the society at large. In the case of the former, capital expenditure is the appropriate mechanism. It allows influential groups to amass wealth through various means, including contracting and real estate. The newly created commerce class remained rich during the bust, but had fewer opportunities to become even richer as spending on massive infrastructure projects dwindled.

As for the rest of the society, the story is different. Making a segment of the society rich is not the same as securing a comfortable standard of living for the society at large. In order to maintain loyalty of the society, the government ensured that subsidies are paid, employment opportunities are available, and free health and education are provided for. Therefore, during the Fourth Development Plan, the government reduced spending on infrastructure by two-third in order to ensure that the government had enough wealth to cover current expenditure. Moreover, during the Fifth Development Plan the government reallocated 34 billion SR from project expenditures to current expenditures.

The changes in the configuration of spending are clearly illustrated in the Fourth Development Plan (1985–1990), the Fifth Development Plan (1990–1995), and the Sixth Development Plan (1995–2000). Whereas current expenditures were slightly cut, capital expenditures were significantly reduced. Table 5.7 provides a breakdown of actual spending on development by sector. The figures in the table show that infrastructure and economic development in all three plans were primary targets of the government cutbacks. For instance, spending on physical infrastructure, including housing, municipalities, and transportation, was reduced from 249 billion SR in the Third Plan to 68 billion SR in the Sixth

TABLE 5.7 *Plan and actual of development expenditures (percentage of total), 1985–2000*

Development plans	Economic development[1]		Human development[2]		Social development[3]		Physical infrastructure[4]	
	Plan	Actual	Plan	Actual	Plan	Actual	Plan	Actual
Fouth plan	26.1	19.9	27.1	35.3	17.9	18.2	18.9	17.6
Fifth plan	15.8	10.6	39.1	47.3	14.7	12.9	27.0	21.0
Sixth plan		11.5		51.5		20.8		16.2

(1) Includes projects related to water, agriculture, industry, trade, services.
(2) Includes projects related to the general education system.
(3) Includes projects related to health, social security, social rehabilitation, etc.
(4) Includes projects related to transportation, telecommunications, and housing.
SOURCE: MINISTRY OF ECONOMY AND PLANNING, FIFTH DEVELOPMENT PLAN (1975–1980), 25; SIXTH DEVELOPMENT PLAN (1980–1985), 65; SEVENTH DEVELOPMENT PLAN (1985–1990).

Plan (see table 5.8). Similarly, spending on economic development was reduced from 262 billion SR in the third plan to 48 billion SR in the Sixth Plan. Because of these drastic declines in economic development and infrastructure, many jobs were lost. As a response the government enacted several initiatives. First, it initiated a "Saudiization" program, which aimed to replace foreign workers with Saudi workers. Second, it authorized the enlargement of employment opportunity in the public sector in order to offset the shortage of employment in the private sector. Third, to stimulate the private sector, government contracts were granted to local instead of foreign suppliers.

Moreover, except for a small reduction in the Fourth Plan, human development jumped from 130 billion SR in the Third Plan to 217 billion SR in the Sixth Plan, constituting the largest share of spending on development. Although not substantially amplified, social development increased from 61 billion SR in the Third Plan to 88 billion SR in the Sixth Plan, ranking as the second largest of government allocations to development. Funding for health, education, and other welfare services continued, and even increased, despite the financial constraints.

Overall, non-capital subsidies were reduced in some years more than others, but were not entirely eliminated. As shown in table 5.9, subsides did not steadily decline but fluctuated, in accordance with changes in oil revenues. In fact, some subsidies increased, such as social security. In actuality, a comparison between the boom and bust shows that differences in the cutback of subsidies

TABLE 5.8 *Actual spending on development (Billion SR)*

Sector	Third plan	Fouth plan	Fifth plan	Sixth plan
Economic development[1]	262	65	35	48
Human development[2]	130	114	155	217
Social development[3]	61	59	63	88
Physical infrastructure[4]	249	87	74	68
Total	702	325	327	421

(1) Includes projects related to water, agriculture, industry, trade, services.
(2) Includes projects related to the general education system.
(3) Includes projects related to health, social security, social rehabilitation, etc.
(4) Includes projects related to transportation, telecommunications, and housing.
SOURCE: MINISTRY OF ECONOMY AND PLANNING, FOURTH DEVELOPMENT PLAN (1985–1990), FIFTH DEVELOPMENT PLAN (1990–1995), AND SIXTH DEVELOPMENT PLAN (1995–2000).

were selective, not uniform, and not always substantial. The largest cutback was in the late 1980s (1989–1992), where subsidies were around 3 billion SR. This, however, was not much different than the amount of subsides disbursed in the boom years between 1977 and 1979. Another sample from the years of boom and bust shows similar behavior. For instance, if one compares the years of 1983, 1984, and 1985 when prices were around $28 a barrel to 1986, 1987, 1988 when oil prices were at least 10 to 15 less, the amount of subsidies disbursed is roughly the same – about 6 billion SR to 8 billion SR. The obvious exception is only in 1981 and 1982, when total subsidies reached 10 billion SR. Capital subsidies, however, were drastically decreased (see table 5.10), but not completely suspended. Even during the worst years of oil revenues the government continued to disburse loans on the magnitude of no less than 4 billion SR. In comparison with boom period, however, there is no doubt that capital loans were much lower. From 1986 to 1999, capital subsidies totaled 75 million SR, about one-third of the amount spent between 1970 and 1985.

Internal Balancing

Despite the decline in oil revenues, military expenditure was not significantly reduced. Spending on defense remained a top priority for the government,

TABLE 5.9 *Government subsidies (Million SR), 1986–1999*

Year	Food[1]	Agricultural	Social security	Electrical	Other[2]	Total
1986	5173	594	1352	1536	666	9321
1987	6724	334	1427	1400	712	10597
1988	10563	304	1396	157	460	12881
1989	5215	282	1380	210	741	7827
1990	5673	199	1315	374	736	8297
1991	6340	82	1323	451	780	8976
1992	5662	723	1333	451	589	8757
1993	5861	283	2514	451	571	9680
1994	5953	368	2654	210	707	9892
1995	3897	23	2573	210	656	7358
1996	2558	336	2693	210	642	6440
1997	2020	267	2891	746	661	6584
1998	1248	276	2998	210	632	5363
1999	1081	247	2990	210	587	5116

(1) Imported and local barley, local wheat, and poultry feed subsidies are included.
(2) Includes social affairs, social care, youth clubs, public transport and others.
SOURCE: MINISTRY OF ECONOMY AND PLANNING, ACHIEVEMENTS PLAN, 2000.

especially after the Gulf War in 1990–1991. According to Grimmett, from 1993 to 1999, Saudi Arabia arms-transfer agreements amounted to $24.5 billion, making Saudi Arabia the leading spender on arms in all developing countries. Out of the $165.2 billion of arms-transfer agreements of developing nations, Saudi Arabia's share was 14.8 percent.[8]

As far as the total amount spent on defense, table 5.11 shows that defense spending fluctuated during the 1980s and 1990s. But overall, allocation to defense did not drastically change and remained above one-third of the government current expenditures. The figures in table 5.11 present the percentage change in defense spending from year to year, as well as defense spending as percentage of total government current consumption. As shown in the table, the share of defense as percentage of total government current consumption decreased. The decline was largely due to an increase in total government current expenditures, not necessarily a significant decline in military spending.

8 Richard F. Grimmett, "Conventional Arms Transfers to Developing Nations, 1993 to 2000," *Mediterranean Quarterly*, no. 13 (Spring 2002): 36–55.

TABLE 5.10 *Subsidized government loans (Million SR), 1970–1999.*

Year	Saudi Arabian Agricultural Bank[1]	Saudi Credit Bank	Specialist Funding Programs[2]	Public Investment Fund[3]	Saudi Industrial Development Fund[4]	Real Estate Development Fund	Total
1986	1,019	240	82	130	355	4,114	5,940
1987	841	268	85	382	543	3,972	6,091
1988	755	293	37	130	439	3,282	4,936
1989	854	323	22	40	829	2,976	5,044
1990	1,017	286	5	364	664	2,446	4,782
1991	757	289	101	44	1,032	1,905	4,128
1992	775	288	64	20	1,179	1,542	3,868
1993	941	345	34	6	1,055	3,147	5,528
1994	671	293	55	0	1,322	4,762	7,103
1995	413	296	73	118	2,001	2,187	5,088
1996	431	306	99	300	2,002	2,401	5,539
1997	627	337	46	542	2,003	2,228	5,783
1998	897	300	87	1,072	1,931	1,633	5,920
1999	903	310	115	542	1,246	1,700	4,816

(1) Includes loans for working capital
(2) Includes loans to contractors, Private hospitals and dispensaries, hotels, newspapers, bakeries etc.
(3) Including equity investments
(4) Includes industrial, electricity, cold storage and dates packing.

SOURCE: MINISTRY OF ECONOMY AND PLANNING, ACHIEVEMENTS PLAN, 2000.

TABLE 5.11 *Allocation to Defense as Percentage of Government Consumption, 1980–2000*

Year	Defense expenditure		Total gov. consumption (Million)	Defense as percentage of gov. consumption
	Million	Percentage change		
1980	43,000	5	81,914	52
1981	57,858	35	1,28,527	45
1982	48,492	-16	1,26,904	38
1983	51,839	7	1,21,625	43
1984	50,330	-3	1,15,647	44
1985	47,236	-6	1,20,051	39
1986	41,392	-12	1,11,633	37
1987	41,739	1	1,13,040	37
1988	36,474	-13	1,02,240	36
1989	42,325	16	1,19,958	35
1990	45,590	8	1,27,824	36
1991	59,195	30	1,69,128	35
1992	56,359	-5	1,52,692	37
1993	51,296	-9	1,30,976	39
1994	45,503	-11	1,22,552	37
1995	46,018	1	1,25,923	37
1996	52,708	15	1,44,783	36
1997	59,618	13	1,61,795	37
1998	55,747	-6	1,55,192	36
1999	42,285	-24	1,54,095	27
2000	51,357	21	1,83,804	28

SOURCE: SAUDI ARABIAN MONETARY AGENCY (SAMA), ANNUAL REPORTS, VARIOUS YEARS.

As for the percentage change from year-to-year in defense spending, the figures in table 5.11 show that out of the 15 years, the government cut spending seven times. The largest three cuts were made when oil prices were extremely low. In 1986, after oil prices plunged to $13 a barrel, the defense budget was reduced from 47 million SR to 41 million SR in 1985 (12 percent). Its share, however, of government consumption remained as high as 36 percent of total consumption. Similarly, in 1988 spending on defense was reduced by 13 percent, which brought down the military spending to about 36 billion SR, its lowest level since 1979. The following year, however, the policy was reversed. In 1989

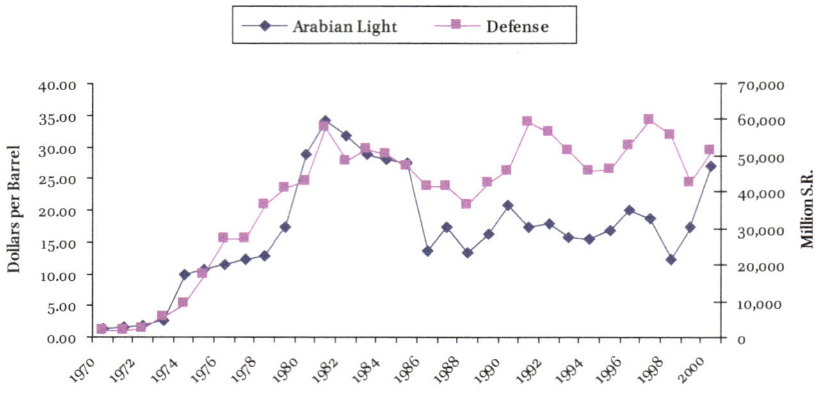

FIGURE 5.2 *Oil price (Arabian Light) vs. defense expenditures.*
SOURCE: SAUDI ARABIAN MONETARY AGENCY, TABLES OF ANNUAL REPORT

military spending increased by 16 percent, which restored military spending to a level of 42 billion SR. The last major cutback was made in 1999 after oil prices reached $12 a barrel in 1998. The cutback in terms of annual percentage change was the largest in the last two decades, 24 percent reduction. But this reduction has two qualifications. First, defense spending in 1997 was nearly 60 billion SR, the highest amount ever spent on defense. Therefore, the two successive reductions, 1998 and 1999 amounting to 30 percent, were made against a record high allocation. Therefore, despite the two successive reductions, the allocation to defense remained significant. Moreover, the two successive reductions were reversed in 2000 by a 21 percent increase. In sum, throughout the 1980s and 1990s, despite low oil prices, defense was a major spending item for the government (see figure 5.2).

Cases of Major Threats

Saddam Hussein's Iraq and the Invasion of Kuwait

In the 1990s the Saudi regime confronted Saddam Hussein's Iraq. Saddam Hussein reasoned that Iraq sacrificed monumentally to face the threat of the Persians, and therefore it was first among equals in the Arab world. However, his pragmatic foreign policy, which lasted for two years, did not yield the intended results: the Arab leaders did not regard him as a vanguard of the Arab world. On top of that, Iraq's oil revenue was not sufficient to deal with the economic problems at home, nor to meet Iraq's tens of billions of foreign debts. Amidst these economic problems, Iraq's relations with Western countries also deteriorated.[9] Moreover, Saddam Hussein became convinced of a conspiracy against Iraq and his regime, prompting him to reorient his policy away from pragmatism. By early 1990s, a confrontational orientation of Saddam Hussein's policy started to emerge with significant security implications for the Saudi regime. Saddam Hussein's belligerent policy initially singled out Western countries, especially the United States and Israel, as the main plotters against Iraq. In early January, Saddam Hussein remarked: "if Israel carries out an act of aggression, the United States will not be able to deny the charge of encouraging Israel to carry out such as attack."[10] Having identified the United States and Israel as the main enemies not only to Iraq, but also to the rest of the whole Arab world, Saddam Hussein called on the Arab world to assist Iraq against Western threats. In his words,

> The group as a whole must encourage those with greater ability to move forward with all their support and good wishes, without letting this lead the more capable ones into adventurism or isolation.[11]

Pan-Arab security became the main theme of Saddam Hussein's speeches, which demanded from the Arab states to rely on each other, and most importantly on the powerful ones, for security of the region. The combination of anti-Western and pan-Arabism sentiment, couched in military rhetoric, was alarming for the Saudi regime. Saudi Arabia maintained a close relationship with the Western countries, especially the United States. Economic consideration and security guarantees made the United States an invincible partner. Pan-Arabism, which was purposely deployed by Saddam for strengthening his regime at the cost of other Arab states, was reminiscent of Nasser's encroachment on the legitimacy of the Saudi regime and a destabilizing force within the kingdom.

The increasing Iraqi hostility toward Kuwait and the United Arab Emirates (UAE), which are members of the Gulf Security Council (GCC), increased Saudi Arabia's suspicion of Saddam Hussein. The occupation of Kuwait shook the Arab world and intensified Saudi Arabia's sense of vulnerability with the pres-

9 Relations between Hussein and Western countries had deteriorated in the wake of the execution of the British journalist, Farzad Bazoft, Iraq obtainment of the Kryton switches required for the detonation of a nuclear weapon, and the British interception of "supergun" parts destined to Iraq.

10 Quoted in Murden, *Emergent Regional Powers and International Relations in the Gulf: 1988–1991*, 134.

11 Quoted in Ibid., 140.

ence of the Iraqi forces in Kuwait. With the liberation of Kuwait the immediate danger of Iraqi invasion of Saudi Arabia was removed. But, having demonstrated its belligerence and indifference to Arab norms, Saddam Hussein's Iraq remained a source of threat. In the aftermath of the war, Saddam Hussein was deeply resentful of the Saudi regime. For its part, the Saudi royal family was also deeply concerned about the outlook of Iraq, if Saddam Hussein were to be ousted. Because of religious fragmentation between Sunni and Shia, Iraq was vulnerable to external manipulation and intervention. Saudi Arabia was particularly concerned that Iran would exploit its religious affiliation with Iraq's large Shi`a community to expand its influence in the Gulf region.

Responses: Internal and External Balancing

Because of the military nature of the threat, the Saudi regime responded with multiple strategies. On the domestic front, the regime increased allocations to the military. In the years after the Iraqi invasion of Kuwait, spending on defense considerably increased. The average amount of spending on defense before the Iraqi invasion was 41 billion SR (1986–1990), but after the invasion the average increased to 51 billion SR (1991–2000). On the external front, the immediate response of the Saudi state was to seek American military support. Four days after the Iraqi invasion, Dick Cheney, then-Secretary of Defense, arrived in Riyadh on a mission to discuss security arrangements, including dispatching American troops. Hosting American troops, however, was not a small concern for the Saudi regime; mainly for fear that it might trigger strong domestic opposition to the royal family. King Fahd, however, was far more concerned about the military threat of Saddam Hussein. Therefore, on the next day it was announced that American troops and planes would be dispatched to the kingdom for the purpose of defense against an Iraqi invasion. The initial deployment consisted of 2,300 troops, but eventually the number of American troops in the kingdom was increased to approximately 430,000.

Meanwhile, the Iraqi invasion precipitated an oil crisis as a result of the disruption of oil supplies from Iraq and Kuwait. It led to the removal of about 4 mbd from the oil market.[12] In response to the panic created by the war and absence of supplies from two major oil producers, the oil price doubled in less than two months. Figures in table 5.12 show the fluctuation in OPEC Reference

12 Three days after the Iraqi invasion of Kuwait the United States and the European community placed a full boycott on oil from Iraq and Kuwait, and a day later Japan took a similar step. Opinion Editorial, "The Oil Weapon," *Washington Post*, August 05, 1990; David E. Sanger, "The Price Confronting Iraq in the Oil Market; Japanese Joining in Embargo Despite Impact on Economy," *New York Times*, August 06, 1990.

TABLE 5.12 *Spot OPEC reference basket price and its components (US dollars per barrel), 1990–1991*

Month-Year	Iraq Basrah Light	Kuwait Export	Arabic Light	OPEC reference basket
Jan–90	16.0	17.3	18.3	20.0
Feb–90	16.0	16.4	17.6	19.0
Mar–90	16.7	15.4	16.5	17.7
Apr–90	15.0	13.6	14.6	15.6
May–90	14.5	13.6	14.6	15.5
Jun–90	12.8	12.0	13.1	14.0
Jul–90	15.5	13.8	15.4	16.3
Aug–90	–	22.2	25.8	26.4
Sep–90	–	–	30.1	32.1
Oct–90	–	–	31.9	34.3
Nov–90	–	–	28.5	30.8
Dec–90	–	–	24.2	26.2
Jan–91	–	–	20.7	22.3
Feb–91	–	–	15.3	17.5

SOURCE: OPEC STATISTICAL BULLETIN, VARIOUS YEARS.

Basket price before and after the invasion. The price climbed from the pre-war level of $16.26 a barrel in July to $32.10 a barrel in September, just after Saddam Hussein invaded Kuwait in August. One of the most vital roles played by Saudi Arabia during and after the Iraqi invasion was to moderate oil prices.

To make up for shortage of supplies and exert downward pressure on oil prices, Saudi Arabia initially sought to increase oil production under the cover of OPEC in order not to provoke Iraq. However, when Libya, Iran, and Algeria balked, Saudi Arabia finally moved unilaterally. On August 19, 1990, the Saudi Oil Minister, Hisham Nazer, announced that his country would increase oil production by 2 mbd with or without the approval of other OPEC members.[13] Saudi Arabia increased the production in the 1990s to an average of one-third of total OPEC production (see table 5.13). Meanwhile, Iraq considered the Saudi policy of higher production and lower price as "an act of aggression." Iraq was

13 Youssef M. Ibrahim, "Confrontation in the Gulf Foreigners in Iraq will Share in Hardships; Baghdad Warns; Saudis to Increase Oil Output; Challenge to OPEC," *New York Times*, August 19, 1990.

TABLE 5.13 *Saudi Arabia oil production (1,000 b/d), 1989–2000*

Country	1989	1990	1991	1992	1993	1994	1995	1996	1997	1998	1999	2000
Saudi Arabia	5,065	6,413	8,118	8,332	8,048	8,049	8,023	8,102	8,012	8,280	7,565	8,095
OPEC	21,138	22,781	23,056	24,715	25,076	25,532	25,588	25,826	26,527	28,820	27,311	28,873
% OPEC	24	28	35	34	32	32	31	31	30	29	28	28

SOURCE: OPEC STATISTICAL BULLETIN, 2005.

TABLE 5.14 *Saudi Arabia oil exports by region (Million barrels)*

Region	1989	1990	1991	1992	1993	1994	1995	1996	1997	1998	1999	2000
North America	380	481	664	615	488	521	504	491	489	544	534	577
W. Europe	321	380	623	636	628	602	598	531	591	646	454	484
Asia	389	593	862	958	986	957	1,006	1,031	1,011	971	922	1,045

SOURCE: OPEC STATISTICAL BULLETIN, 2005.

not only defying the military, but also intended to use higher oil prices as a pressure against Western powers and to hinder war efforts against Iraq. However, pressed for revenues and determined to shield Western countries from high prices, Saudi Arabia expanded its output. For instance, as shown in table 5.14, oil exports to North America, Western European, and Asia and Pacific continued to steadily increase.

In addition to oil supplies, petrodollars were a crucial balancing mechanism. Unlike conventional balancing in which armed forces are pooled, Saudi Arabia had little to offer the United States and other European powers in terms of military support, but had plenty of petrodollars. Hence, money was used to strengthen the coalition against Iraq and to pay for war efforts. To win Russia's support, Saudi Arabia approved $4 billion of credit line to the financially strapped Russian government. And to help ease the financial hardship caused by higher energy prices, Saudi Arabia provided $800 million to Eastern Europe.[14] At the same time, the kingdom contributed about $400 million to 500 million a month to the daily costs of American troops, and paid as well for much of the transportation, equipment, and daily costs of many Third World troops.[15] Ultimately, according to one estimate, the kingdom spent about $62 billion.[16]

14 Rachel Bronson, *Thicker Than Oil: America's Uneasy Partnership with Saudi Arabia*, (Oxford University Press, 2006), 197–198.

15 Chronology, *Middle East Journal* (Winter 1991), 87.

16 Nadia El-Shazly and Raymond Hinnebush, "The Challenge of Security in the Post-Gulf War Middle East System," in Raymond Hinnebusch and Anoushiravan Ehteshami, eds., *The Foreign Policies of Middle East States* (Colorado: Lynne Rienner, 2002), 81.

Saudi Arabia also sought support from other Arab and Muslim countries. Although American troops enforced the defense capability of the kingdom, the regime could not ignore domestic opposition to the hosting of non-Muslim troops in the Muslim holy land. To legitimatize the presence of American troops in the kingdom, the kingdom made two attempts. First, it approached the highest religious authority in the kingdom, the Grand Mufti, Bin Baz, to sanction the deployment of the American troops. Second, it sought and received support from other Arab and Muslim countries, such as Egypt and Turkey. Mubarak's Egypt accommodated the needs of the royal family.[17] It played a key role in the organization of an Arab League emergency summit on August 10 in Cairo, and for the active diplomacy behind obtaining 12 votes out of 20 attended members, which supported a resolution that condemned the Iraqi invasion and denounced the Iraqi force buildup on Saudi Arabia's border. The resolution provided the legitimate political cover for Saudi Arabia to request military support from the United States against a possible Iraqi attack.[18] The resolution set off a rally of criticism against using force and inviting non-Muslim/non-Arab troops to Saudi Arabia. Mubarak countered these claims by sending Egyptian troops to the kingdom, deploying 17,500 soldiers. In return for Mubarak's support, Saudi Arabia agreed to write off Egyptian debt and provided Egypt with aid of approximately $500 million. Moreover, Saudi Arabia in concert with the United Arab Emirates and Qatar agreed to revive the AOI, the Egypt-based Arab military-industrial complex.[19] Likewise, Turkey was rewarded with $800 million and the kingdom pledged another $1 billion for a Turkish special defense fund.[20]

17 Arab countries that did not support Saudi Arabia were financially punished. Jordan, for example, did not support the use of international force. King Hussein of Jordan declared that the Kuwaiti crisis was "an Arab problem," and declined to send Jordanian troops to Saudi Arabia until Western troops pulled out. Consequently Saudi's oil shipment and financial aid to Jordan was suspended. Similarly, Saudi-Yemen relations deteriorated, because Yemen did not support a military action against Iraq. The response of the Saudis was to financially squeeze Yemen, by withholding financial aid and canceling work privileges of Yemenis working in Saudi Arabia. In the end, hundreds of thousands of Yemenis were deported to Yemen.

18 Ann Mosely Lesch, "Contrasting Reactions to the Persian Gulf Crisis: Egypt, Syria, Jordan, and the Palestinians," *Middle East Journal* (Winter 1991): 30.

19 The AOI was set up in 1975 as a military-industrial joint venture between Egypt and the Arab oil-producing countries. As a response to Egypt participation in the Camp David Accords in 1978, along with other Arab oil-producing countries, Saudi Arabia halted financial support, bankrupting the AOI.

20 Bronson, *Thicker Than Oil: America's Uneasy Partnership with Saudi Arabia*, 197–198.

Islamist Opposition[21]

The decision of the royal family to invite American troops to the kingdom (discussed above) precipitated a furious discontent among Islamist groups. Although After the Gulf War Islamist opposition to the regime became a long-standing security problem.[22] Saudi Arabia has long been opposed to the idea of hosting American troops, fearing opposition against the royal family. Ultimately the regime decided to risk a confrontation with Islamists, rather than risking a larger and most likely far grimmer military confrontation with Saddam Hussein.

Still, the Islamist threat of the 1990s was unmatched by any other internal challenge the kingdom had faced since the *Ikhwān* rebellion in 1929.[23] Unlike the Marxist and Nasserite threats in the 1950s and 1960s, because of the Islamic nature of the Saudi society, the message of Islamists enjoyed a wider appeal among major elements of the society, especially among the youth.[24] And unlike the 1970s threat, which emanated from the Shia community in the Eastern province, the Islamic opposition of the 1990s was broadly Sunni-based, with leaders and followers from Najid (central Arabia), the traditional powerhouse of the Saudi ruling family. Even the uprising of Juhaiman Al Utaibi and his followers in 1979, which seized the Grand Mosque in Mecca, was not seriously threatening because Al Utaibi was not backed by a popular support. On the other hand, capitalizing on a popular view of the royal family as corrupt and subservient to the United States, this wave of Islamist opposition struck alliances with different groups within Saudi society and penetrated state institutions and bureaucracy.[25]

---

21 By Islamist opposition I am broadly referring to various Islamist groups, including reformist, rejectionist, and Jihadist. Although these groups have different agendas and were shaped by different circumstances, they stand on a common ground in their opposition to Al-Saud. For more information on these groups, see International Crisis Group, "Saudi Arabia Backgrounder: Who are the Islamists?" Middle East Report Number 31 (September 2004).

22 The presence of American troops in the Kingdom was not the only reason that led to an awakening of Islamic radicalism and militancy; the Israeli-Palestinian conflict and the perception of the Al-Saud as a pro-Western regime also fueled the anger of the Islamic opposition.

23 The *Ikhwān* affair was discussed in chapter two of this dissertation; see pp. 32–34.

24 `Abdulaziz O. Sager, "Political Opposition in Saudi Arabia," in Paul Aarts and Gerd Nonneman, eds., *Saudi Arabia in the Balance: Political Economy, Society, Foreign Affairs* (London: Hurst & Company, 2005), 253.

25 Mamoun Fandy, "Serious Threats to Saudi Arabia's Stability," *Christina Science Monitor*, no. 88 (January 19, 1996), 19.

In the immediate years following the Gulf War, Islamist opposition groups begun to publicly air out their views. Petitions were submitted to King Fahd. In May 1991 the "Letter of Demands" was presented with 400 signatories, including Islamic scholars, university professors, and prominent preachers. In September 1992, 107 religious scholars prepared the "memorandum of Advice." In 1993, Mohammed al-Massari established the Committee for the Defense of Legitimate Rights (CDLR). Prominent preachers like Salman al-Awda, Safar al-Hawali, Ayidh al-Qarni, and Nasir al-Omar became known for their zealous sermons, "denouncing the state's failure to conform to Islamic values, corruption, and subservience to the U.S."[26]

By the mid-1990s the relationship between the Islamists and the Saudi regime took a fatal turn. In November 1995, a car bomb exploded near an American military building in Riyadh, killing five Americans and two Indians and injuring 60 others. In June 1996, a second bomb exploded targeting the Al-khobar military housing complex in the city of Dhahran, killing 19 American airmen and injuring hundreds of others. The attacks were allegedly linked to Iran, but neither the Saudi nor the United States government proved these allegations with concrete evidence.[27] As a result of these attacks and threats of future attacks, the State Department authorized the departure of dependents of American military servicemen and civilian employees.[28]

Responses: Internal Validation and Internal Balancing

The regime moved to address two key issues, the perceived lack of loyalty of citizens and inadequate capabilities of the Saudi military, both deemed serious threats to security. Thus, it was pronounced in the Sixth Development Plan: "To continue supporting and enhancing the on-going development of the Kingdom's own defense capabilities, and to deepen citizens' loyalty and sense of belonging."[29] As concern over citizens' loyalty heightened, the royal family increased spending and intensified internal security. It increased allocations to both human and social developments in the Sixth Plan. The share of human development of total

26 See International Crisis Group, "Saudi Arabia Backgrounder: Who are the Islamists?"

27 Gause makes the same conclusion, "The general trends in both Iranian policy and in Gulf-Iranian relations provide strong circumstantial evidence that Iranian involvement in the domestic politics of the Gulf Monarchies has, if not ended, been greatly reduced from the level of the 1980s." Gregory F. Gause III, *Political Opposition in the Gulf Monarchies* (Florence: European University Institute Working Papers, 2000), 14.

28 AP, "Americans Given Option Of Leaving Saudi Arabia," *New York Times*, July 23, 1996; David Stout, "U.S., Citing Security, Shuts Saudi Embassy Briefly," *New York Times*, October 7, 1998.

29 The Kingdom of Saudi Arabia, Ministry of Economy and Planning, *Seventh Development Plan*, Chap. 2.

development allocation increased from 155 million SR in the Fifth Development Plan to 217 million SR in the Sixth Development Plan (see table 5.8). Also, allocation to social development went up. Similarly, current account expenditures, entailing wages, compensations, and defense expenses, were also increased. Internal security was tightened and a massive arrest campaign was undertaken in which hundreds of Islamists were arrested.[30] The regime increased allocations to security, as shown in table 5.11.[31]

Conclusion

By the mid-1980s the global demand for oil had de-accelerated, and supply had outpaced demand. As a result, the balance of power shifted away from the producers toward the consumers. Thus, oil rentier states, including Saudi Arabia, became asymmetrically dependent on consuming countries. By the same token, the vast amount of oil resources of the kingdom provided the Saudi regime with some influence over other rentier states.

Under such pattern of dependence, the Saudi regime continued with the dual policy of cooptation through internal validation and control through internal balancing, although at lower intensity than during the boom era. Internal validation proceeded with the rolling out of three development plans: the Fourth Development Plan (1985–1990), the Fifth Development Plan (1990–1995), and the Sixth Development Plan (1995–2000). The primary objective of these plans was not different from the previous plans – to raise living standards and create employment opportunities. Likewise, military allocation declined in the period from 1986 to 1990. It increased during the Gulf War, but afterward fluctuated until the end of the decade. To finance the budgetary deficits the state used different options. On the expenditures side, allocations were reshuffled in order to preserve public spending on key social welfare programs. On the revenues side, the government adopted different measures: increased reliance oil exports, turned to domestic debts, exhausted foreign reserves, and took foreign debt.

30 Youssef M. Ibrahim, "Saudi Arabia Arrests 110 Muslim Militants," *New York Times*, September 28, 1994; Youssef M. Ibrahim, "Saudi Arabia Cracks Down on Islamic Militants, Seizing Many," *New York Times*, September 22, 1994.

31 In 1992 seven members were dismissed from the Supreme Authority of Senior Scholars, the highest-level religious body in the kingdom, for declining to support the government. See Youssef M. Ibrahim, "Saudi King Ousts 7 Senior Clerics For Acts of Criticism by Omission," *New York Times*, December 15, 1992. Prince Sultan, the Minister of Defense, was appointed to head the newly created Supreme Council of Islamic Affairs.

In the case of the Iraqi invasion of Kuwait, the Saudi regime adopted a policy of external balancing. It cooperated with the United States and other Western powers to repulse Saddam Hussein from Iraq. While Western powers, especially the United States, employed military force against Iraq, Saudi Arabia resorted to oil power. It helped to calm the oil market, by making up for the missing Kuwaiti and Iraqi supplies. In addition, the regime used petrodollars to help finance war efforts against Saddam Hussein, providing billions of dollars to the United States as well as financial aid to other Arab and Muslims countries that dispatched troops to the kingdom. In addition, the Saudi regime relations with Islamists worsened as a result of Gulf War. The Saudi government responded by increasing spending on human and social development as well as security.

CHAPTER 6

The Second Boom (2000–2012)

The oil market recovered, and prices surged to a record high during this period. With a growing demand in developing countries and higher oil prices, Saudi Arabia's fiscal budget significantly improved, allowing the kingdom to increase spending on security and self-defense. As in previous chapters, the first section examines the global patterns of oil supply and demand as well as the changes in oil reserves and supplies of Saudi Arabia. Section two focuses on the security strategies of internal validation and balancing. The last section reviews the importance of oil resources in Saudi Arabia's responses to specific security challenges.

Pattern and Scope of Dependence in the Boom Era (2000–2012)

Oil Market: Supply and Demand

Global demand increased by 0.15 mbd in 2001 and 2.9 mbd in 2002. The attacks on September 11, 2001, worsened an already weak global economy. Demand picked up in 2003 and continued through 2004, with growth rates of 2.17 percent and 3.28 percent, respectively. In 2005, demand deteriorated. Slowing OECD economies caused much of the decline. OECD countries experienced successive decline from 2007 to 2012 (see table 6.1). In contrast with the period 1985–2000, in which demand increased by 19 percent, between 2004 and 2007, demand stagnated at 50 mbd. The financial crisis of 2008 plunged Western economies into the worst recession since the Great Depression. As a result, world demand declined by 0.3 mbd in 2008. The largest reduction took place in the major oil consumers, OECD countries, where demand declined by 1.6 mbd. Consumption in OECD countries dropped further by 2 mbd in 2009. On the other hand, economic growth in Asian and other developing countries, in particular China and India, strengthened the demand for oil. In 2005, developing countries accounted for almost 75 percent of oil demand growth. India alone accounted for 30 percent of the non-OECD Asia group.[1] After the financial crisis, economic growth in non-OECD countries continued to drive the strong demand for oil.

---

1 OPEC, OPEC Annual Report 2005.

 | DOI 10.1163/9789004277731_008

TABLE 6.1 *World oil consumption (Million b/d), 2000–2012*

Year	Total OECD	Total Non-OECD	World
2000	48	29	77
2001	49	29	77
2002	49	30	78
2003	49	30	80
2004	50	32	82
2005	50	34	84
2006	50	35	85
2007	50	36	86
2008	48	36	85
2009	46	38	85
2010	47	40	87
2011	46	42	88
2012	46	43	89

SOURCE: EIA, ANNUAL ENERGY REVIEW.

On the supply side, Middle Eastern production produced on average more than the previous boom and bust eras, averaging 23 mbd, with the lowest production in 2002 of 18 mbd and a peak production of 24 mbd in 2012. In comparison with other regions, Middle Eastern countries remained the most important suppliers, producing about one-third of the world's total production (see table 6.2). As in previous decades, Latin American production maintained its expansion of supply. Second to the Middle East, production from Latin America remained consistently between 9 mbd and 10 mbd, accounting for 13 percent to 14 percent of the world's total production. Venezuela and Mexico were the biggest producers, together producing more than half of the total production from Latin America. Production gradually expanded from African countries, increasing from 6.7 mbd in 2000 to 2008 mbd in 2012. Production patterns from Eastern and Western Europe differed from those in previous decades. In Western Europe, production reversed growth from previous decades and experienced a steady decline, with its share of total production dropping from 9.54 percent in 2000 to 3.9 percent in 2012. Unlike Western Europe, Eastern Europe reversed the decline of the previous decade, increasing production from 7.6 mbd in 2000 to 12.6 mbd in 2012.

TABLE 6.2 *Oil production by region (1000 b/d), 2000–2012*

Year	North America	Latin America	Eastern Europe	Western Europe	Saudi Arabia	Africa	Asia and Pacific	Middle East[1]
2000	7,213	9,317	7,631	6,288	8,095	6,746	7,253	13,316
2001	7,179	9,327	8,250	6,034	7,889	6,613	7,208	12,888
2002	7,191	9,475	9,040	5,952	7,093	6,429	7,276	11,525
2003	7,140	9,549	9,961	5,628	8,410	7,246	7,288	11,998
2004	6,824	9,962	10,746	5,375	8,897	8,277	7,347	13,085
2005	6,538	10,130	11,083	4,905	9,353	8,816	7,446	13,369
2006	6,448	10,078	11,532	4,502	9,208	8,958	7,311	13,679
2007	6,499	9,796	11,997	4,320	8,816	9,066	7,309	13,679
2008	6,349	9,635	12,046	4,047	9,198	9,191	7,414	13,944
2009	6,570	9,495	12,396	3,829	8,184	8,461	7,346	12,685
2010	6,705	9,654	12,651	3,528	8,166	8,656	7,597	12,865
2011	6,913	9,813	12,650	3,192	9,311	7,431	7,458	13,694
2012	7,813	9,687	12,670	2,890	9,763	8,203	7,463	14,370

SOURCE: OPEC, ANNUAL STATISTICAL BULLETIN, 2012.

Saudi Arabia: Oil Reserves and Exports

By the end of 2012, Saudi Arabia had produced more than 211 trillion barrels of oil since 1960. At the same time, in the period of 2000–2012, there was little increase in its proven reserves. Its proven reserves in 2010 were comparable to the levels of the 1960s, almost 25 percent less than in the year 2000. Still, as shown in table 6.3, relative to other regions, the kingdom possessed an enormous share of proven reserves.

Although Saudi Arabia's oil exports substantially increased over the bust period (1986–2000), its share of the international oil market slightly fluctuated between 2001 and 2012. Its share dropped to about 14 percent of world export in 2002 and went back up to 19 percent in 2012. For the rest of the years, it hovered around 16 percent. In comparison with the 1990s, Saudi Arabia's share of total exports experienced a steady and gradual decline. The relative decline of Saudi Arabia's share of world export is partly because of the substantial growth in oil exports from Eastern European countries. Starting at the end of the 1990s, Eastern European countries have substantially increased their exports. As shown in table 6.4, Eastern Europe exports doubled in the period 2000–2007. Helped by new technology and development of new oil fields, Russian

TABLE 6.3 *Proven oil reserves by region as percentage of total world reserves, 2000–2012*

Region/Country	2000	2005	2010	2011	2012
North America	2.5	3.2	2.7	1.9	1.9
Latin America	11.2	10.4	22.8	23.0	22.9
Eastern Europe	8.7	10.9	8.7	8.0	8.1
Western Europe	1.7	1.3	0.9	0.8	0.8
Saudi Arabia	24.1	22.2	18.0	18.1	18.0
Africa	8.6	9.9	8.9	8.7	8.8
Asia and Pacific	3.6	3.3	3.0	3.2	3.5
Middle East-SA	39.6	40.2	36.1	36.3	36.1

SOURCE: OPEC, ANNUAL STATISTICAL BULLETIN, 2012.

production significantly increased during 2001–2012. Its net export increased from 4.1 mbd in 2000 to 2006.2 mbd in 2009. By far, Russia was the biggest producer in Eastern Europe, exporting between 4 mbd and 5 mbd. Another factor is the soft growth in demand from OECD countries. Still, Saudi Arabia remained the largest oil producer and exporter, with a unique ready capacity of increasing oil production to 12 mbd and a potential increase to 15 mbd.

Patterns of Dependence during 2000–2012

Except for 2009, oil-consuming countries were asymmetrically dependent on producers. From 2001 to 2004, global demand grew by almost 5 mbd. Even when prices started to surge in 2005, oil demand continued to rise, reaching an increase of 1.4 mbd over the previous year. In 2005, oil WTI prices jumped to a new high of $56.51 a barrel. Rising prices and strong demand in non-OECD countries continued in subsequent years. In 2007, despite prices rising from $50s a barrel to low $90s a barrel in 2007, demand increased over the previous year by 818,000 barrels a day. Although oil prices went up from $93 a barrel in January 2008 to $134 a barrel in July 2008, global demand expanded in the first two quarters of 2008.

On the other hand, supply fell short in keeping up with the demand. From 2002 to 2008, the oil market experienced a number of supply disruptions. The first disruption took place when Venezuela state-owned oil company Petróleos de Venezuela S.A. (PDVSA) went on strike in December 2002. The strike removed 3 mbd from the market. Efforts by OPEC to make up for the Venezuelan oil losses caused a decline in the global spare capacity. The second major disruption is

TABLE 6.4 *Oil exports by region (1000 b/d), 2000–2012*

Year	North America	Latin America	Eastern Europe	Western Europe	Saudi Arabia	Africa	Asia and Pacific	Middle East[1]
2000	1,231	4,906	4,145	4,991	6,253	5,209	2,230	9,835
2001	1,168	4,817	4,594	4,883	6,036	5,098	2,275	9,196
2002	1,175	4,478	5,584	4,793	5,285	5,127	2,333	8,515
2003	1,252	4,523	6,482	4,419	6,523	5,763	2,095	8,069
2004	1,376	4,606	7,309	4,262	6,813	6,380	2,006	9,573
2005	1,378	4,739	7,727	3,649	7,209	6,482	1,961	9,686
2006	1,393	4,900	8,199	3,765	7,029	6,583	1,877	9,915
2007	1,422	5,164	8,553	3,919	6,962	6,883	1,923	10,006
2008	1,564	4,479	5,658	2,786	7,322	6,397	1,743	10,253
2009	1,535	4,271	6,230	2,775	6,268	6,771	1,613	9,230
2010	1,431	4,203	5,632	2,451	6,644	6,613	1,506	9,678
2011	1,735	4,566	5,809	2,248	7,218	5,355	1,399	10,524
2012	1,817	4,625	5,886	2,113	7,557	6,182	1,384	10,889

(1) Saudi Arabia is not included.

SOURCE: OPEC, ANNUAL STATISTICAL BULLETIN, 2012.

the US-led invasion of Iraq in March 2003. In the months following the invasion, Iraqi production fell to 1.3 mbd from the prewar level of 2.8 mbd.[2] Insecurity and frequent attacks on oil infrastructure hindered resumption of oil production to prewar level. Production improved but remained unstable due to lack of security. The third disruption was caused by hurricanes Katrina and Rita, which effectively shut down oil and gas production from the Gulf of Mexico. In 2004, the Outer Continental Shelf accounted for 27 percent of the total U.S. production. The storms caused major damage to infrastructure, including the destruction of 111 production platforms and a number of drilling rigs.[3] As a result, from June 2005 to June 2006, an average of 450,000 barrels a day was lost from the

2 K. Kirabaeva, F. Parodi, and Y. Said, "Iraq Selected Issues," (April 26, 2013). IMF Country Report No. 13/218. Accessed August 2014. http://www.imf.org/external/pubs/ft/scr/2013/cr13218.pdf. In the years to come, Iraq continued to produce less than its target of 2.5 mbd. On average, its production reached 2,011 in 2005 and 1,878 in 2005.

3 Lawrence Kumins and Robert Bamberger, "Oil and Gas Disruption From Hurricanes Katrina and Rita," CRS Report for Congress, RL 33124. Accessed September 2014. http://fpc.state.gov/documents/organization/55824.pdf.

market.[4] In 2007, another major producer, Russia, reduced supply by 117,000 barrels a day from the previous year's level. Overall, in 2005 and 2008, non-OPEC supply declined by 203,000 barrels a day and 185,000 barrels a day, respectively, while OPEC supply declined in 2006 and 2007.

As a result of the global financial crisis and higher production from Saudi Arabia, the WTI oil price per barrel plunged from mid-$130 a barrel in July to low $40s in December 2008. During this period of low oil prices, the influence of the kingdom increased with respect to other rentier states. Empowered by its high reserves and large spare capacity, Saudi Arabia had the capability to pressure prices further downward. Unlike other countries, Saudi Arabia was less vulnerable to lower prices because of its unique capacity to produce more oil and to partly compensate for any revenue losses.

Under these conditions, consuming countries were asymmetrically dependent on producing countries between 2001 and 2012. This means Saudi Arabia was in a strong financial position to spend more money on internal as well as external balancing and validation. At the same time, the tight market reduced the influence of Saudi Arabia with respect to other rentier states. Table 6.5 shows the financial situation of the kingdom during this period. Oil revenues declined in 2001 and continued to deteriorate in 2002, dropping by more than 22% from its level in 2000. The Iraq invasion led to the loss of more than 2 mbd of Iraqi oil. To make up for the Iraqi oil loss, Saudi Arabia and other OPEC members increased their production. Saudi Arabia's exports jumped from 5.2 mbd in 2002 to 6.5 mbd in 2003. The combination of higher production and oil prices increased Saudi oil revenues by 39 percent. Oil revenues continued to rise because of higher prices and production until the last quarter of 2008. Higher production from Saudi Arabia in the first three quarters of 2008 combined with the onset of the financial crisis in the last quarter of 2008 led oil prices to plunge by the end of the year. As a result, Saudi oil revenue declined by 549 trillion SR in 2009 or a 56 percent decline from its level in 2008. In 2010, revenues increased by 54 percent. Rising oil prices and higher production earned Saudi Arabia revenues worth an all-time high of 1,144 trillion SR in 2012.

Internal Validation

Unlike the bust era of 1985–2000, higher prices in this decade enriched the kingdom's coffers. As shown in table 6.6, except for the years 2002, 2003, and 2009, government budget registered surpluses. Those surpluses were mainly higher in the second part of the decade when oil prices were the highest. Total

4 EIA, " STEO Supplement: Why are oil prices so high?" Accessed July 2014. http://www.eia.gov/forecasts/steo/special/pdf/high-oil-price.pdf.

TABLE 6.5 *Saudi Arabia oil revenue, 2000–2012*

Year	Oil revenues (Million SR)	Oil revenue as percentage of total revenue	Oil price Arabian Light	Oil export (1,000 b/d)
2000	214,424	0.8	26.8	6,253
2001	183,915	0.8	23.1	6,036
2002	166,100	0.8	24.3	5,285
2003	231,000	0.8	27.7	6,523
2004	330,000	0.8	34.5	6,813
2005	504,540	0.9	50.2	7,209
2006	604,470	0.9	61.1	7,029
2007	562,186	0.9	68.8	6,962
2008	983,369	0.9	95.2	7,322
2009	434,420	0.9	61.4	6,268
2010	670,265	0.9	77.8	6,644
2011	1,034,360	0.9	107.8	7,218
2012	1,144,818	0.9	110.2	7,557

SOURCE: SAUDI ARABIAN MONETARY AGENCY (SAMA), TABLES OF ANNUAL REPORT; FOR OIL EXPORTS: OPEC, ANNUAL STATISTICAL BULLETIN, 2006 & 2012.

TABLE 6.6 *Budgetary deficit and surplus (Million SR), 2000–2012*

Year	Total revenues	Total expenditure	Deficit/surplus
2000	258,065	235,322	22,743
2001	228,159	255,140	-26,981
2002	213,000	233,500	-20,500
2003	293,000	257,000	36,000
2004	392,291	285,200	107,091
2005	564,335	346,474	217,861
2006	673,682	393,322	280,360
2007	642,800	466,248	176,552
2008	1,100,993	520,069	580,924
2009	509,805	596,434	-86,629
2010	741,616	653,885	87,731
2011	1,117,792	826,700	291,092
2012	1,247,398	873,305	374,093

SOURCE: SAUDI ARABIAN MONETARY AGENCY (SAMA), ANNUAL REPORTS, VARIOUS YEARS.

revenues were highest in 2008 when oil prices reached a record high of more than $130 a barrel. Consequently, total revenues reached more than a trillion SR. Under such favorable financial conditions, government expenditures tripled over the decade, with greater spending allocated to current rather than capital expenditures.[5] Figure 6.1 shows the allocation patterns of current and capital expenditures. As in previous years, the difference in spending between current and capital was largest when oil prices were low. In 2000, when oil prices were in the mid-$20s, current spending amounted to 12 times the capital spending. In 2008, when oil prices averaged $77 a barrel, current expenditure was about two times the capital expenditure. Generally speaking, capital expenditure diminished at times of lower oil prices. As oil prices started to rise in 2005, both current and capital expenditures significantly increased. Current expenditure increased from 217 billion SR in 2000 to 2284 billion SR in 2005 and then to 612 billion SR in 2012. Similarly, capital expenditure increased from 18 billion SR in 2000 to 2062 million SR in 2005 to 2262 billion SR in 2012.

As in previous years, the government channeled spending through development plans: the Seventh Development Plan (2000–2004) and the Eighth Development Plan (2005–2010). In both plans, the government accorded spending priority to human resources development, creation of employment opportunities, and improving living standards by offering free health and education and

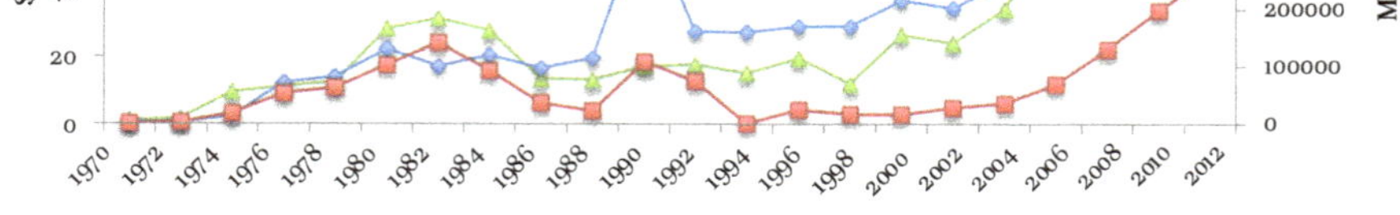

FIGURE 6.1 *Oil price vs. Current and capital expenditures*
SOURCE: SAUDI ARABIAN MONETARY AGENCY (SAMA), ANNUAL REPORTS, VARIOUS YEARS.

5 Capital spending includes construction of roads, schools, and hospitals and acquiring new equipment, while current spending includes expenditures on goods and services consumed in the current year, such as defense, wages, subsidies, and welfare programs. As explained in the previous chapter, the Saudi government prioritizes capital spending due to its immediate impact on domestic stability. See chapter 8, pp. 132–133.

various subsidies and loans. The Seventh Development Plan was constrained by moderate oil prices. In comparison with the Sixth Development Plan, spending in the Seventh Plan did not dramatically change. Total spending increased from 421 billion SR in the Sixth Plan to 485.3 billion SR in the Seventh Plan. The pattern of spending also remained the same. As shown in table 6.7, human development received the largest share, followed by social, infrastructure, and then economic development. In total, 276.9 billion SR was allocated to human resources, representing 57.1 percent of total expenditures. Spurred by higher oil prices, the government almost doubled expenditures on all sectors during the Eighth Development Plan. In total, government expenditures reached 864 billion SR. Human resources accounted for 55.6% of total expenditures, amounting to an increase by 42%. Similarly, while social development increased by 40 percent, physical infrastructure and economic development doubled.

Subsidies and loans remained important in the Seventh and Eighth Development Plans. As shown in table 6.8, noncapital subsidies did not decline in the first part of the decade despite relatively lower oil prices. Expenditure on subsidies jumped from 5 billion SR in 1999 to more than 6 billion SR in 2000, thereafter drastically increased. Total spending reached 10 billion SR in 2004 and then increased by almost threefold, amounting to 28 billion SR, in 2010. The total amount spent from 2000 to 2010 reached 151 billion SR. In comparison with the boom and bust eras, total subsidies spent during this period exceeded the bust era (1986–1999) by 39 percent and the boom era (1970–1985) by 76%. The total value of subsidies more than doubled during the Eighth Development Plan, when oil prices were relatively higher than in previous years. In fact, the annual

TABLE 6.7 *Actual spending on development (Billion SR)*

Sector	Seventh plan	Eighth plan
Economic development[1]	54.4	105.8
Human development[2]	276.9	480
Social development[3]	92.6	155.9
Physical infrastructure[4]	61.4	122.2
Total	485.3	863.9

(1) Includes projects related to water, agriculture, industry, trade, services
(2) Includes projects related to the general education system.
(3) Includes projects related to health, social security, social rehabilitation, etc.
(4) Includes projects related to transportation, telecommunications, and housing.
SOURCE: MINISTRY OF ECONOMY AND PLANNING, SEVENTH DEVELOPMENT PLAN (2000–2005) AND EIGHTH DEVELOPMENT PLAN (2000–2010).

TABLE 6.8 *Government subsidies (Million SR), 2000–2012*

Year	Food[1]	Agricultural[2]	Social security	Others[3]	Total
2000	2,210	267	2,983	837	6,366
2001	2,069	287	3,038	836	6,231
2002	2,766	242	3,012	839	6,859
2003	4,909	302	2,986	931	9,128
2004	5,981	296	3,022	1,010	10,310
2005	3,168	300	3,011	1,082	7,560
2006	2,850	308	6,922	1,302	11,382
2007	4,092	83	10,768	1,781	16,723
2008	9,747	83	13,663	1,689	25,181
2009	8,212	91	13,043	2,260	23,606
2010	5,165	93	15,230	7,811	28,298
2011	–	–	–	–	–
2012	–	–	–	–	–

(1) Imported and local barley, local wheat, and poultry feed subsidies are included
(2) Includes subsidies paid by Agricultural Development Fund and Ministry of Agriculture and Water
(3) Includes social affairs, social care, youth clubs, public transport and others.
SOURCE: MINISTRY OF ECONOMY AND PLANNING, ACHIEVEMENTS PLAN, 2000.

subsidies spent in the years 2006 to 2010 exceeded any other amount disbursed during previous decades. In 2000, food subsidies doubled but fluctuated throughout 2012. It peaked in 2008, reaching 9.7 billion SR, and afterward declined to 5 billion SR in 2010. Agricultural subsidies diminished due to a shift in policy away from large-scale crop production to preserve scarce water resources.[6] The largest increases were allocated to social security and social-related affairs. In the period 2000–2010, the value of social-related affairs increased by 10 times, while social security by 5 times.

Capital subsidies gradually increased during the Seventh and Eighth Development Plans. The total amount of loans extended by government credit institutions jumped from 317 billion SR in 2003 to 2456 billion SR in 2010. The Real Estate Development Fund (REDF) made the largest contribution, facilitating the development of personal and commercial real estate development. REDF gradually increased the value of loans for the construction of private houses

6 For example, the government reduced subsidies for growing wheat and barley.

and investment, reaching 172 billion SR in 2010, representing in that year 36.1 percent of the total value of loans extended by government credit institutions. Second to REDF is the Public Investment Fund (PIF). The value of its loan to public and quasi-public enterprises increased significantly during the Eighth Plan, reaching 15.5 billion SR in 2009. The value of its loans throughout 2010 amounted to 116.2 billion SR. The Saudi Industrial Development Fund (SIDF) and the Agricultural Development Fund (ADF) continued to provide significant amounts of loans. Up to 2010, SIDF extended loans at a total value of 97.5 billion SR, while ADF provided loans worth 42.3 billion SR.

Internal Balancing

According to the Stockholm International Peace Research Institute (SIPRI), Saudi Arabia ranked seventh and eighth among the top 15 military spenders in 2011 and 2012.[7] Defense spending continued to exceed other expenditures, averaging 27 percent of total spending in the period 2001–2012. During this period, Saudi military spending exceeded military spending during the bust period (1985–2000) by 162 billion SR.[8] Between 2000 and 2005, allocation to the military was comparable to the 1990s, with an average spending of 52 billion SR. The share of defense as percentage of total government current consumption decreased because of an increase in total government current expenditures. In absolute terms, government spending increased dramatically. The figures in table 6.9 show the percentage year-to-year changes. Between 2002 and 2012, the government cut spending only twice. The first time in 2002, after oil prices plunged to $13 a barrel, the defense budget was reduced from 47 million SR in 1985 to 2041 million SR. However, its share of government consumption remained as high as 36 percent of total consumption. On the heels of the financial crisis, the government cut spending the second time in 2009. Spending was reduced by 13 percent from 2008, amounting to a reduction of 12 billion SR. However, higher oil prices in the following year allowed for recovering military spending to its level prior to the financial crisis. Thereafter, between 2001 and 2012, higher oil prices boosted military spending, which increased at a rate of 12 percent. In 2012, military spending reached 120 trillion SR, more than double the level of military spending in 2000 (see Figure 6.2).

---

7 Stockholm International Peace Research Institute (SIPRI), "Trends in World Military Expenditures, 2013," Accessed August 2014. http://books.sipri.org/product_info?c_product_id=476.

8 Saudi Arabia does not report spending on internal security separately from their military spending.

TABLE 6.9 *Allocation to defense as percentage of government consumption, 2000–2012*

Year	Defense expenditure		Total gov. consumption	Defesne as percentage of gov. consumption
	Million	Percentage change		
2000	51,357	21	183,804	28
2001	51,696	1	188,695	27
2002	48,652	-6	184,517	26
2003	51,773	6	198,148	26
2004	58,669	13	221,798	26
2005	63,949	9	262,650	24
2006	75,206	18	311,082	24
2007	80,157	7	322,086	25
2008	97,754	22	345,098	28
2009	85,632	-12	357,015	24
2010	96,010	12	375,968	26
2011	107527	12	488,062	22
2012	120582	12	542,699	22

SOURCE: SAUDI ARABIAN MONETARY AGENCY (SAMA), ANNUAL REPORTS, 2000–2012.

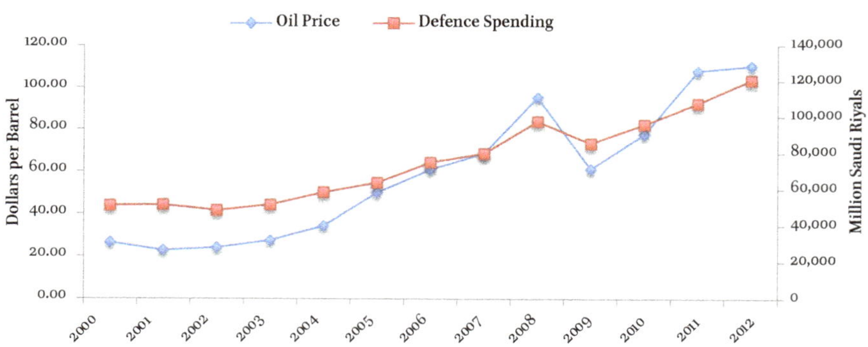

FIGURE 6.2 *Oil price (Arabian Light) vs. defense expenditures*
SOURCE: SAUDI ARABIAN MONETARY AGENCY (SAMA), ANNUAL REPORTS, 2000–2012.

The Iranian Threat

Since the Islamic Iranian Revolution erupted in 1979, Iran-Saudi relation has been colored with tension, rivalry, and confrontation. At the heart of the conflict are differences over ideology, political ambitions, and the role of the US in the Middle East. The self-proclaimed position of Saudi Arabia as defender of the Islamic faith and custodian of the holy places clashed with the ideals and discourse of the Islamic Revolution. Spearheading the revolution, Ayatollah Khomeini leveled strong criticisms against the Saudi royal family and called for its overthrow.

Relations between the two countries worsened with the breakout of the war between Iraq and Iran. Not only were diplomatic relations severed, but Saudi Arabia also sided with Saddam Hussein's Iraq against Iran. It was not until the death of Khomeini that diplomatic relation was restored. With the election of pragmatic leaderships in Iran, the 1990s marked a period of détente between the two countries. During the presidency of Ali Akbar Hashemi Rafsanjani, relations with Saudi Arabia improved, leading to an exchange of visits at the ministerial level as well as in oil policy. The rapprochement continued with the election of Ali Mohammad Khatami as president of Iraq in 1997. During Khatami's presidency, a cooperation agreement was signed with Saudi Arabia in economic, commercial, technical, scientific, cultural, and sports areas.[9]

Still, Iran-Saudi relations remained bogged down with distrust and major differences. They continued to differ on oil policy and the presence of the US in the Gulf. The Saudi regime continued to mistrust Iran's intention in the region and worry about its military capability, including its conventional and nonconventional forces. Following the US-led invasion of Iraq in 2003 and the collapse of Saddam Hussein's Sunni regime, tension increased between Iran and Saudi Arabia over Iraq. Given the Shia majority of Iraq, Saudi Arabia feared that Iran would gain a strong presence in a new Shia-led government in Iraq, which would also expand Iran's influence in the Gulf and the Levant.

With the election of Mahmoud Ahmadinejad, the Iran-Saudi rapprochement came to an end. From the beginning, President Ahmadinejad pursued a

---

9 Anthony Cordesman, *Saudi Arabia Enters the Twenty-First Century: The Political, Foreign Policy, Economic, and Energy Dimension* (Praeger Publishers: Westport, 2003), 44–51. For more information on Iran-Saudi relations see also Frederic Wehrey, Theodore W. Karasik, Alireza Nader, Jeremy Ghez, Lydia Hansell, Robert A. Guffey, *Saudi-Iranian Relations Since the Fall of Saddam: Rivalry, Cooperation, and Implications for U.S. Policy* (Rand National Security Research Security, 2009). Accessed August 2014. http://www.rand.org/pubs/monographs/MG840.html.

nonsectarian hard-line foreign policy, supporting Sunni and Shia groups in the Arab world. Under the leadership of Ahmadinejad, Iran took a stronger pro-Palestinian line, providing financial support to the Sunni Palestinian organization HAMAS. At the same time, President Ahmadinejad continued to provide political and military support to Hezbollah. Finally, Ahmadinejad aggressively supported the creation of a friendly Shia government in Iraq.[10]

Responses: External Balancing

At the height of the tension between Saudi Arabia and Iran, the Saudis turned to oil production and revenue to balance against Iran. Using oil money, Saudi Arabia propped up allies to counter Iranian influence in the Levant. Following the 2006 war between Israel and Hezbollah, Saudi Arabia transferred $1 billion to the Lebanese Central Bank and provided an additional $500 million to support the reconstruction effort. Albeit much less than the Saudi pledge, Iran also donated $120 million to be spent on schools, clinics, and bridges in the Shia surroundings of Beirut. Similarly, Iran aligned with HAMAS, while Saudi Arabia supported the Palestinian Authority (PA). To shore up support for the Palestinian Authority and counter the popularity of HAMAS, Saudi money flowed into the coffer of the PA. In 2009, Saudi Arabia donated $1.1 billion to pay salaries of PA employees and security officers.[11] The Saudis followed the same strategy in financing groups and individuals that could counterbalance, if not roll back, Iran's influence in Iraq. In WikiLeaks cables, US Ambassador Christopher Hill reports the following:

> The Saudis have used considerable financial and media resources to support Sunni political aspirations, exert influence over Sunni tribal groups, and undercut the Shia Islamic Supreme Council of Iraq (ISCI) and Iraq National Alliance.[12]

---

10 For more on Iran's strategy in Iraq, see Mohsen M. Milani, "Iran's Strategies and Objectives in Post-Saddam Iraq," in *Iraq, Its Neighbors, and the United States: Competition, Crisis, and the Reordering of Power*, eds. Henri J. Barkey, Scott B., and Phebe Marr (Washington DC: United States Institute of Peace, 2011).

11 Wehrey, Karasik, Nader, Ghez, Hansell, Guffey, *Saudi-Iranian Relations Since the Fall of Saddam: Rivalry, Cooperation, and Implications for U.S. Policy.*

12 Simon Tisdall, "WikiLeaks Cables: Saudi Arabia Rated a Bigger Threat to Iraqi Stability than Iran," *Guardian*, December 5, 2010.

Given that Iran is highly dependent on oil sales, the Iranian economy and government budget is vulnerable to lower oil prices. In 2006, the *New York Times* reported as follows:

> Cheney and King Abdullah said the king has presented Mr. Cheney with a plan to raise oil production to force down the price, in hopes of causing economic turmoil for Iran without becoming directly involved in a confrontation.[13]

In the last quarter of 2007, the Saudis raised production by 333,000 barrels a day. Despite the production hike, the OPEC basket oil price reached more than $80 a barrel. To pressure the price downward, the Saudis increased production in the first, second, and third quarters of 2008. Production went up from 9,057 mbd in the first quarter to 9,460 mbd in the third quarter. The OPEC basket oil price climbed to $131.22 a barrel in July 2008; however, prices started to decline thereafter. In December 2008, the OPEC basket oil price plunged to $38.60 a barrel. The plunged price was far below what Iran needed to avoid budget difficulties, an oil price above $75 a barrel.[14]

Arab Spring

The so-called Arab Spring posed an unprecedented and complicated challenge to the Saudi royal family. The demonstrations threatened to radically transform state-society relations and alter the regional status quo. At the domestic level, protestors called for political reform and for an end to authoritarian regimes. Calls for democracy, dignity, and accountability were the slogans of protestors. In two countries, Tunisia and Egypt, mass protest overcame the heavy-handed authoritarian states and their security forces. In Tunisia, the birthplace of the Arab revolutions, President Zine El Abidine Ben Ali was forced to flee to Saudi Arabia on January 14, ending more than two decades of an autocratic regime.[15] Egypt was next in line. After two weeks of demonstrations against President Hosni Mubarak, the 18-day revolt achieved its goal. On

---

13 Hassan M. Fattah, "Bickering Saudis Struggle for an Answer to Iran's Rising Influence in the Middle East," *New York Times*, December 22, 2006.

14 Simon Romero, Michael Slackman, and Clifford J. Levy, "Oil-Rich Countries Face a Reckoning," *New York Times*, October 20, 2008.

15 More than two thousand miles away from Saudi Arabia, the Arab Spring started in an unlikely place in the rural part of Tunisia. The self-immolation of a street vendor Mohamed Bouazizi, protesting mistreatment of the authority, sparked a countrywide protest.

February 11, President Mubarak stepped down. The outcome of the uprisings in Egypt and Tunisia signaled two important lessons to Saudi Arabia. First, the power of the people could overwhelm the ironfisted regime and could force political change. Second, the spread of the protests from Tunisia to other Arab countries demonstrated the widespread discontent and appeal of democracy and accountability among the masses in the Arab world. The threat of the domino effect escalated when protests erupted closer to home in Yemen, Bahrain, and then at home in Saudi Arabia.

Inspired by Tunisia and Egypt, Yemenis began to protest in January. After the resignation of President Mubarak of Egypt, jubilant protestors called on President Ali Abdullah Saleh to step down. President Saleh deployed security forces and supporters to confront protestors at Sanaa University Square. On March 18, snipers shot dead at least 50 protestors, promoting a countrywide anger and sorrow. The carnage galvanized support for protestors from government officials, military officers, members of Parliament, and around the country. After 10 months of protest, Saleh's 33-year rule came to an end.[16]

As Saleh's regime teetered on the brink of collapse under the pressure of mass unrest, Saudi Arabia faced two possible undesirable scenarios: Yemen plunging into a civil war or the emergence of a democratic government. Given their long-shared border, a Yemeni civil war implies security spillovers and sure Saudi intervention in Yemen. On other hand, a functioning democratic Yemeni government is unlikely to be receptive to Saudi influence and might inspire calls for more political freedom within Saudi Arabia.

The protest reached the shore of Bahrain in February. Thousands of mostly Shia protestors took to Pearl Square in Manama, calling for political and economic reform.[17] Their grievances of political discrimination and economic exclusion have long been a source of tension with the Sunni-dominated Bahraini state and the Sunni Al Khalifa ruling family. As in previous episodes of demonstrations, clashes erupted as police used force to clear protestors from downtown Manama. The violence left a number of demonstrators dead, prompting angered protestors to call for the end of Al Khalifa's rule. Consequently, the state widened its clampdown on protestors and

16 Sheila Carapico, "No Exit: Yemen's Existential Crisis," in *The Arab Revolts: Dispatches on Militant Democracy in the Middle East*, eds. David McMurray and Amanda Ufheil-Somers (Bloomington: India University Press, 2013).

17 Grievances of the Shia population have been a major source of tension with the Sunni ruling family.

portrayed the demonstrations as an Iranian plot to undermine the Sunni rule.[18]

Two related factors heightened Saudi concern: Saudi Arabia's tense rivalry with Iran and its long-standing tension with the Shia population in the Eastern Province. As discussed below, the Shia population in the Eastern Province harbored grievances of discrimination by the Saudi state and called for more rights and political freedom. For its part, the Saudi government accused Iran of sowing dissent among the Shia population and encouraging rebellion. Starting in February 2011, the Shia in the Eastern Province of Saudi Arabia started to demonstrate, and online activists used Facebook to organize a "Day of Rage" on March 11. The combination of extensive government warning and heavy security presence in major cities deterred protestors from taking to the streets.[19]

Responses: Internal Validation and Internal Balancing

Saudi responses to the Arab uprisings varied in each case. At home, it employed internal validation and balancing. On February 23, King Abdullah bin Abdul Aziz promised to spend $36 billion in social benefits. Less than two months later, the government announced even more enormous spending. The government pledged $66.7 billion for building 500,000 housing units and $4.3 billion for new medical facilities. In addition, the government raised the minimum wage for government employees to $800 and promised an unemployment allowance of $260. Moreover, the package included a bonus worth two months' salary for all government employees and $500 for all university students.[20] On the internal balancing side, the government added 60,000 new jobs to the Interior Ministry.

In the case of Egypt and Yemen, Saudi Arabia stepped in to ensure the survival of the regime. To counter revolutionary forces, the kingdom balanced with embattled Egyptian and Yemeni regimes. In the case of Egypt, while the revolution ousted President Mubarak, the military-based regime remained all-powerful. So the Saudis balanced with the Egyptian military against the Muslim Brotherhood. Within days after the Egyptian military toppled the elected Egyptian President Mohamed Morsi of the Muslim Brotherhood, Saudi Arabia, Kuwait, and the United Arab Emirates pledged $12 billion in aid to Egypt.[21]

---

18 Cortni Kerr and Toby Jones, "A Revolution Paused in Bahrain," in *The Arab Revolts: Dispatches on Militant Democracy in the Middle East*, eds. David McMurray and Amanda Ufheil-Somers (Bloomington: India University Press, 2013).

19 Ian Black, "Saudi Arabian Security Forces Quell 'Day of Rage' Protests," *Guardian*, March 11, 2011.

20 "Saudi monarch announces billions in handouts," accessed July 2014. http://www.aljazeera.com/news/middleeast/2011/03/201131817411791 6648.html.

21 Rod Nordland, "Saudi Arabia Promises to Aid Egypt's Regime," *New York Times*, August 19, 2013.

Similarly in Yemen, the Saudis tried to mediate a solution that would prevent Yemen from descending into civil war or from transitioning to democracy.[22] To avert a civil war scenario, the Saudis pressured President Saleh to step down. At the same time, to counter democratic forces from gaining the upper hand, the Saudis balanced with key players in Saleh's regime. Under the guise of the Gulf Cooperation Council, Saudi Arabia orchestrated the transfer of power from President Saleh to his Vice President Abd-Rabbuh Mansour al-Hadi. In a single-candidate election, Mansour became president in February 2012. Afterward, Saudi Arabia supported the new Yemeni government with $3 billion in fuel and cash in 2012.[23]

In Bahrain, Saudi Arabia took drastic measures to prevent a Shia-led uprising from succeeding in ending the Sunni Al Khalifa rule. From the start of the Bahraini revolution, the Saudi monarchy balanced with the Al Khalifa ruling family. Under the leadership of Saudi Arabia, the Gulf Cooperation Council pledged $20 billion in aid to Oman and Bahrain over a period of 10 years, with each country receiving $10 billion. As the standoff continued between the Sunni regime and the Shia protestors and violence escalated, Saudi Arabia remained firmly in support of Al Khalifa. On March 14, a day after the intense confrontation between police and protestors, Saudi Arabia deployed 1,000 troops to Bahrain in support of Bahrain's ruling family.

Conclusion

With the turn of the 21st century, demand for oil picked up, especially from developing countries. Oil prices dramatically increased, reaching well above $100 a barrel. The tight market implied that consuming countries were asymmetrically dependent on producers, especially the most important producer—Saudi Arabia. Flushed with oil money, the Saudi government could afford to spend at a larger scale on both internal validation and balancing. As in previous decades, internal validation continued through its development strategies, namely, the Seventh Development Plan (2000–2004) and the Eighth Development Plan (2005–2010), lavishly spending on health, education, agriculture, water, and

22 Madawi Al-Rasheed, "Saudi Internal Dilemmas and Regional Reponses to the Arab Uprisings" in *The New Middle East: The Protest and Revolution in the Arab World*, ed. Fawaz Gerges (Cambridge University Press, 2014).

23 "Unable to balance the books," *The Economist*, accessed August 2014. http://www.economist.com/blogs/pomegranate/2014/06/yemens-economy.

various subsidies. At the same time, spending on defense, including internal and external security forces, more than doubled.

During this period, the relationship between Saudi Arabia and Iran reverted to tension and hostility. Saudi Arabia tried to check growing Iranian influence in the Levant by providing financial support to groups and individuals closely aligned with the Saudi regime. Moreover, the Saudi government engineered a collapse of the oil price in 2008. Heavily dependent on oil revenue, Iran benefited from high oil prices, but Saudi Arabia cranked up production throughout 2008 in order to bring down the price and to deprive Iran of lucrative oil revenue. The second important security challenge that confronted the Saudi government was the Arab Spring. From the start, Saudi Arabia attempted to halt and reverse the changes unleashed by the Arab Spring. Balancing by means of providing financial, political, and military support, the kingdom managed to secure friendly ruling regimes in Egypt, Yemen, and Bahrain. At home, the regime resorted to internal validation and balancing to deter mass protests.

CHAPTER 7

Conclusion

Utilizing the political economy approach, this book theorizes the impact of oil relations on security strategies of Saudi Arabia. As an oil rentier state par excellence Saudi Arabia provides a good case to theorize the relationship between different patterns of oil dependence and security. Specifically, the types and intensity of Saudi Arabia's security strategies are examined during in the period 1950–2012, covering oil market conditions of pre-boom, boom, bust, and boom. According to rentier theory, the two important features of rentier states are the externality of state revenues; usually in the form of oil export receipts, and public expenditures as high percentage of GDP. The core political logic of rentier theory claims that in times of oil boom rulers refrain from taxation and indulges in public spending in order to buy allegiance of their society. This simple logic has four primary shortcomings.

First, it does not distinguish between government current and capital expenditures. The difference is crucial: by granting lucrative infrastructure and development contracts, capital expenditure performs as an effective mechanism of increasing the wealth of selective groups, namely members of the royal family, the commercial class, and state elites. On the other hand, current expenditure refers to allocation of social services, subsidies, and public employment. Therefore, in times of budgetary deficit, it is less risky for the government to cut capital than current spending.

Second, rentier theory sidesteps the era of bust. What are the implications of slumped prices for political stability? What are the likely security policies in times of financial crisis? For these questions rentier theory offers no explicit answer. Thus, the analyst is left to presume that the most likely outcome is political instability. Yet, in Saudi Arabia, during the bust period, political crisis did not take place in the wake of a decline in government spending. Third, rentier theory adopts a narrow view of economic resources, limiting oil resources to public spending. However, oil revenues can be used to build powerful internal security and military capabilities. Likewise, under certain conditions, oil supplies can be used as an instrument to influence oil producers and consumers. For instance, the state might embargo supplies to consuming countries in order to boost its image at home. A case in point is the 1973 oil embargo.

Finally, there is no account of international politics in rentier theory, despite the fact that oil supplies and revenues are highly intertwined with the international political economy. Therefore, the omission of international politics is

 | DOI 10.1163/9789004277731_009

particularly puzzling. Consequently, rentier theory is incapable of explaining how domestic and foreign policies are interconnected through relations of oil dependence. Similarly, theories of realism and liberalism were shown to lack explanatory power in the context of the rentier state. The commonly deployed Weberian conception of the state, in which taxation is taken for granted as the primary source of revenues, by default excludes the rentier state from the analysis.

Building on the concept of dependence, some Middle East scholars proposed applying the concept "asymmetric interdependence" to account for Saudi Arabia's external security behavior. Their main shortcoming is the lack of conceptual clarity. Because the logic of dependence is not sufficiently explored in the Middle East literature, its explanatory power has contributed little, if any, in the way of demystifying the relationship between the rentier state and the international political economy. This book fills this gap by combining insights from both disciplines: Middle East area studies and international relations.

In actuality the concepts of dependence and rentier theory are highly related, because the rentier state is embedded in the international political economy. In order to provide a fuller account of rentier state politics, it is necessary to provide a deeper understanding of both concepts: rentier theory and dependence. Although rentier theory is underdeveloped in various ways, it offers a powerful framework for conceptualizing a deviant case of Weberian states. To enhance the explanatory power of rentier theory, assumptions and insights from realism and liberalism are incorporated. In doing so, I developed a new approach called neo-rentier theory.

Neo-rentier theory is based on two realist assumptions: security is the dominant objective of the state, and relative military capability is an effective means of ensuring security. But it departs from realism in two ways. First, the conception of security is broadened to include internal security. Second, the conception of power is broadened to include not only relative military capability, but also the concept of economic dependence. Without disclaiming the essentiality of military power, the theory demonstrates under what conditions dependence serves as a source of power and influence, and it specifies causal mechanisms underlying different security strategies. Drawing on the liberal tradition, economic dependence is defined in terms of opportunity costs for actors involved in the relationship. To conceive dependence as a source of power for actor A over actor B, it implies that opportunity costs are low for A but high for B. In other words, severing or altering the relationship causes hardship for B but not for A. If both actors have high opportunity costs, then the relationship is characterized as mutual or symmetric dependence.

The model hypothesizes that different patterns of dependence generate constraints on and opportunities for a rentier state, influencing availability and intensity of the following security strategies: internal validation, internal balancing, external validation, and external balancing. On one hand, oil revenue can be used for financing internal validation, internal and external balancing; on the other hand, oil supplies and prices can be used for external balancing and external validation. Different patterns of dependence between a given rentier state and other states (consumers and producers) not only affect the amount of accrued oil revenues, but also the possibility of manipulating supplies and price.

When the rentier state is asymmetrically dependent on other actors (oil consumers and producers), strategies of internal validation and internal balancing can be employed at a low intensity while external validation or external balancing is less accessible. When other actors (oil consumers and producers) are asymmetrically dependent on a rentier state, strategies of internal validation and internal balancing can be employed at a higher intensity while external validation or external balancing is more accessible.

In what follows, I describe the patterns of dependence that existed with consumers and producers in each phase, and the strategies adopted by the Saudi Arabian state in response to internal and external threats. Lastly, I discuss what lies ahead for Saudi Arabia's security policies in the coming two decades.

The Pre-Boom Era, 1950–1970

Despite the growing demand for oil in the aftermath of the Second World War, Saudi Arabia was hardly in a position to influence supplies and prices, for two reasons. First, the levers of control over oil resources were in the hands of the oil companies (ARAMCO), not the Saudi state. Since oil was discovered in Saudi Arabia in 1938, until the early 1970s, the state had no active role in deciding on issues related to expanding production or determining oil prices. ARAMCO had full control over oil decision-making, while the kingdom was merely a receiver of oil royalties and profit sharing. Such income was a primary source of state income, providing an amount of 70 percent to 88 percent of total revenues (see table 3.5, Chapter 3). Second, during the 1950s and 1960s supplies of oil were plentiful in the world. Despite worldwide growing demand for oil in the non-communist world, vast supplies were available from North and Latin America (see table 3.2, Chapter 3). Consequently, despite its considerable reserves and production capacity, the Saudi Arabian state was unable to exploit its oil resources for influence vis-à-vis consumers and other oil producers. On the other hand, during this period, the Saudi Arabian state was heavily dependent on income from ARAMCO (and by extension the consuming countries). In other words, opportunity costs were low for oil consumers but high for Saudi Arabia. In this period (1950–1970), therefore, Saudi Arabia was dependent on oil consumers and had no relations of dependence with other oil producers.

During this period, the kingdom employed minimal internal validation and internal balancing in comparison to later periods. The arrival of oil revenues facilitated public spending on a higher scale and wider range of issues than in previous decades. The largest portion of spending was allocated to the development of armed forces and the internal security organization. Improvement in revenues also helped the state to set up new institutions for overseeing the development of various social programs and infrastructure projects, including health, education, and communication and transportation. However, the scope of these programs was limited by the financial terms of the arrangement made with ARAMCO.

In response to Nasser's pan-Arabism in the period 1958–1962, the Saudi regime responded by employing an appeasement strategy. Despite the fact that Arabism embodied ideas that threatened the legitimacy of the Saudi monarchy, the Saudi regime settled for a policy of courting rather than confrontation with Nasser. Both King Saud and Crown Prince Faisal expressed some form of support for Nasser's pan-Arabism. In addition, the kingdom under the leadership of Faisal suppressed its connection with the United States and downgraded its relation with anti-Nasser Arab regimes, Lebanon and Jordan.

The embargos of 1956 and 1967 show that the Saudi regime had no choice but to stop the flow of oil in order to mitigate domestic uproar and avoid isolation in the Arab world. While both cases of oil disruption challenged the consuming states and oil companies to come up with efficient and cost-effective schemes of sharing and rerouting oil supplies, neither embargo produced an energy crisis. In the case of the 1956 embargo, shortages begun to develop, but Western Europe, including the targeted countries, Britain and France, did not plunge into an energy crisis. Partly the oil shortage was attributed to the Arab embargo, but also partly due to the refusal of President Dwight Eisenhower to authorize emergency oil supplies until both countries withdrew their forces. A month after the Canal closed and with a promise from France and Britain to pull out their troops, supplies from the United States resumed. By the early months of 1957, the oil shortages came to an end.

From the very start of the 1967 embargo, the United States came to the aid of Western nations. Its production increased by 1 mbd. Other countries (Venezuela, Indonesia, and Iran) also increased production. Moreover, the availability of supertankers made it easier and faster to transport oil around the globe. As a result, the embargo did not create an oil crisis in Western countries. However, the embargo proved to be costly for the Arab oil-producing countries, including Saudi Arabia, who had to forgo substantial revenues. In sum, although Saudi Arabia used the strategy of external validation, its success was limited because alternative supplies were readily available to Western countries from other oil producers.

With the outbreak of the Yemen Civil war and the deployment of Egyptian troops to Yemen, Nasser's political hostility transformed into a military threat. The proximity of Yemen and the military capabilities of the Egyptian troops heightened Saudi concern. The Saudi regime's response took several forms. First and foremost, the regime abandoned the appeasement strategy. At home, King Faisal initiated strategies of internal validation and internal balancing. As for validation, the regime sought to boost support at home with the ten-point reform and development program. With regard to internal balancing, the regime paid particular attention to internal security apparatuses. It increased allocations to the Ministry of Information, Ministry of Interior, Intelligence Bureau, and the National Guards.[1] Abroad, it employed external balancing on several fronts. However, oil resources alone did not determine all these balancing acts. First, Faisal activated an alignment policy based on Islamic ideology. The purpose of this strategy was to thwart the spread of Nasser's Arab socialism at home and neighboring countries, but also to create an Islamic alliance. Second, the Saudi regime balanced with royalists against the republicans backed by Nasser. This balancing act of Saudi Arabia entailed mainly providing financial aid and arms.

Finally, Saudi Arabia sought to ally with the United States against Nasser's Egypt. The fact that American consumers were not dependent on Saudi Arabia yet becomes clear in the failed attempt of Faisal to acquire unequivocal support from the United States against Nasser. Its inability to more fully exploit oil dependence left the kingdom vulnerable to Nasser's subversion and military threats. Without any power over oil prices or the ability to influence supplies, Saudi Arabia could not extract the desired support from the United States against Egypt.

The Boom Era, 1970–1985

The structural changes in the oil market that took place in the 1970s and early 1980s created the conditions for oil resources to play a more prominent role in

1 Between 1960 and 1970, allocations to the National Guard increased by 325 percent and the Ministry of Interior by 4570 percent. The Ministry of Information and Intelligence Bureau were created in 1965, with a budget of 75 million SR and 15 million SR, respectively. In 1970 allocations were increased to 99 million SR for the Ministry of Information and 18 million SR for the Intelligence Bureau. See table 3.8, Chapter 3.

the internal and external security strategies of Saudi Arabia than in previous decades. By the early 1970s the two constraints of the previous decades had disappeared: first, world demand outpaced world supply. In large part, the transformation of the United States from a net exporter to a net importer intensified demand for oil. Because its own proven reserves had been declining at a rate greater than its discovery rate of new reserves, oil imports rapidly climbed. By the early 1970s the United States became the largest net importer in the world, importing annually an average of 8.5 mbd. Second, the Saudi Arabian state took control over oil decision-making from ARAMCO. The takeover put the Saudi state in charge of a vast amount of proven reserves and a massive export capacity. The combined effect of these two factors, a tight oil market and state control of oil resources, created a situation in which consumers were asymmetrically dependent on Saudi Arabia (as well as on other oil producers).

By 1981, however, the market started to turn against producers, as oil glut began to develop in the market. As a result, rentier states became vulnerable to downward pressure on oil prices. Helped with its vast reserves and production capacity, Saudi Arabia was in a relatively stronger position than other rentier states to cope with lower prices. In sum, during the boom, both consumers and oil rentier states were asymmetrically dependent on Saudi Arabia. Domestically, the strategies of the previous decades were intensified. As a result of the drastic jump in export volumes and oil prices, the state dramatically increased spending on internal validation and internal balancing. In terms of validation, new policies were instituted which substantially enlarged welfare services, consumption and production subsidies, and development projects.[2] In actuality, these policies not only aimed to buy societal acquiescence but also facilitated wealth accumulation for the dominant interest groups. In addition to the awesome magnitude of spending on internal validation, considerable capital was allocated to internal balancing. A vast amount of wealth was allocated to the development of military and internal security capabilities, making the kingdom the leading spender on arms in the region.[3]

---

2 For instance, government expenditure during the Third Development Plan (1980–1985) reached 783 billion SR, which is 751 billion SR more than what the government spent during the First Development Plan (1970–1975). Likewise, in the period 1970–1985, government subsidies increased from 40 million SR to 6,155 million SR. For more information on allocations to development, subsidies, and public employment see tables 4.6–4.9, Chapter 4.

3 Allocation to internal security apparatuses (the Ministry of Information, National Guard, Intelligence Bureau, and the Ministry of Interior) increased from 888 million SR in 1970 to 2033 billion SR in 1984. The Ministry of Defense, however, took the largest allocation, consuming alone in the period 1970–1985 about 70 percent of the total security budget. See table 4.10, Chapter 4.

In the international arena, the kingdom acquired the capability to exert political pressure on consumers and other producers far beyond its own military capabilities, using external validation and external balancing. The four threat cases demonstrated the ability of exercising strategies of external validation and external balancing with respect to both consuming and producing countries. The cases of the October 1973 War and the 1978 Camp David Accords demonstrate the usage of external validation, while the cases of the Iranian Revolution and the Iran-Iraq (1980–1988) War shows the usage of external balancing, through petrodollars and oil supplies.

The events of the 1973 oil embargo demonstrated the kingdom's influence over consuming states. Acting in concert with other Arab oil producers, the kingdom employed cutbacks in volumes and a full embargo in order to influence the position of Western consuming countries on the Arab-Israeli conflict. Unlike the previous embargos of 1956 and 1967 – although the 1973 embargo did not roll back the occupation of the 1967 war – the consuming countries confronted a severe energy crisis and economic difficulties. By contrast, Saudi Arabia and other oil-producing countries were less hurt by the reduction in oil supplies, because their losses in volumes were compensated by higher oil prices.

Likewise, albeit less drastic, is the case of the Camp David Accords. The fallout of the agreement caused a rift in the Arab world and introduced a new source of threat and hostility to the Saudi regime. The Arab leaders and public opinion were fully against Camp David. Under these conditions, the Saudi regime sought to serve its security interest by adopting a policy of external validation, in which the relationship with the United States was suppressed and the relationship with Egypt was broken off. As a demonstration of its disillusion with the United States and solidarity with the Arab cause, the kingdom decided to abandon its role as a price moderator. It announced a ceiling of 8.5 mbd, almost 2 mbd below its oil production level of September 1978. In an already tight market, the reduction had an escalating effect on prices. As a result, prices on the spot market reached as high as $23 a barrel. With a net import of over 10 mbd, the United States was more than ever before sensitive to high oil prices. As for Egypt, all kinds of financial aid were terminated, including credits and the suspension of the AOI.

With respect to the producing countries, the shift from a market driven by demand to an oil glut created asymmetric dependence between Saudi Arabia and other rentier states. As shown in the case of the Iran-Iraq war, the Saudi state exploited asymmetric dependence for security purposes. The case of the Iran-Iraq War demonstrates the usage of oil supplies and revenues as means for external balancing. Considering that the military and ideological threat of revolutionary Iran was perceived to be more immediate than that of Iraq, Saudi Arabia sought to prevent an Iranian victory. Its effort to support Iraq included not only providing financial support, but also depriving Iran from oil revenues. At a critical turn in the course of the war, when Iran was in the process of mounting a massive offense against Iraq, Saudi Arabia enacted a policy of "net-back," where the official oil price was to be calculated as the final-product price less refining and transportation costs. The kingdom also raised oil exports from 2 mbd in 1985 to a little over 3 mbd in October 1986.[4] By tying oil prices to the market rate, instead of charging fixed spot prices, and increasing production, Saudi Arabia effectively allowed the prices to fall: the oil price fell bellow $10 a barrel. The collapse in oil prices had a dramatic budgetary effect on Iran, halving oil revenues in 1986. In sum, from the early 1970s until the mid-1980s both oil consumers and producers became dependent on Saudi Arabia. As a result, the relations of asymmetric dependence affected Saudi Arabia's security behavior at home and abroad. It provided greater resources for internal validation and internal balancing, and facilitated the deployment of external validation and external balancing vis-à-vis consumers and producers.

The Bust Era, 1985–2000

As the bust phase unfolded, the relationship between consumers and producers shifted in favor of oil-consuming countries. The oil glut lasted throughout the 1990s (see table 5.1, Chapter 5). Weak world demand and excess world supply made the opportunity costs of severing or altering the relationship too costly for the producers, but not for the consumers. This was evident by the UN embargo on Iraqi oil exports. Despite the removal of 2 mbd of Iraqi oil, the existence of surplus capacity prevented a surge in oil prices. The absence of Iraqi oil was offset by increased production from other oil producers. The oil glut had a second implication for rentier states: considerable decline in revenue. With low oil prices and because of its considerable oil reserves and export capacity, Saudi Arabia remained powerful with respect to other producers. Therefore, Saudi Arabia could employ external validation and external balancing. At the same time, by definition, as a rentier state, Saudi Arabia was dependent on oil sales. It follows that Saudi Arabia was asymmetrically dependent on oil consumers.

---

4 It was reported that Saudi Arabia aimed to raise production closer to its OPEC quota, 4.3 million barrels. Fred Marc Biddle, "The Saudi Strategy In Cutting Oil Prices," Chicago Tribune, September 22, 1985.

Because of its dependence on the consumers, not only was Saudi Arabia no longer able to use external validation and external balancing, but it had to modify its spending on internal validation and internal balancing. In the case of validation, the government reduced allocations to infrastructure projects and economic development, including transportation, agriculture, service, and trade. Subsidies too were lowered, especially in the case of government loans to the public and private sectors (i.e., capital subsidies). On the other hand, more spending was allocated to services related to social and human development, such as health, education, and social security. Moreover, the state remained unwilling to introduce income tax. In other words, internal validation was not eliminated, but reduced to a level that was unlikely to provoke political instability.[5]

Similarly, internal balancing was also affected by declining oil prices (see table 5.11, Chapter 5). However, when comparing military expenditures in the 1990s against the late 1970s and early 1980s, the cutback is insignificant. For instance, in 1981, when the oil price hit $34 a barrel and the budget had a surplus of 83 trillion SR, allocation to defense amounted to 61 trillion SR (in constant 1980 SR). In 1994, when the oil price sank to $15 a barrel and the budget turned to a deficit of 34 trillion SR, spending on the military reached 47 trillion SR (in constant 1980 SR).

To finance the budget, the government used multiple options, including increasing production volume, exhausting foreign reserves, and borrowing from domestic and international sources. Moreover, the government reduced public spending. While capital expenditure decreased, current expenditure did not. Although capital expenditure is more important for the economy in the long term, current expenditure is more important to buy off political support in the short term. This finding suggests two noteworthy points. First, oil price is not a reliable indicator of political stability in rentier states. Second, the indiscriminate use of public expenditures to draw conclusions regarding political stability, as often done by Middle East scholars, is problematic, as it overlooks the political importance of current expenditures versus capital expenditures.

The case of the Gulf War demonstrates how Saudi Arabia employed strategy of external balancing against producers. When Iraq invaded Kuwait, the Saudi

5 The Saudi government reduced spending on development projects and subsidies. For instance, spending during the Fourth Development Plan (1985–1990) was near 325 billion SR, about half the amount spent during the Third Development Plan (1980–1985). Government subsidies, especially non-capital subsidies, were drastically reduced from 11 million SR in 1985 to 1986 million SR in 1986. See tables 5.8–5.10, Chapter 5.

royal family felt threatened of Saddam Hussein's intension and capabilities. As a result, the Saudi regime sought and received military protection from the United States. Almost 43,000 American troops were deployed to the kingdom. Although the arrival of the American troops enforced the military defense capability of the kingdom against the potential threat from Iraq, it raised major internal security concerns. Islamists strongly opposed the presence of non-Muslim troops in the kingdom. To counter such criticism, the Saudi regime invited troops from other Arab and Muslim countries, including Egypt and Pakistan.

In light of Iraq's annexation of Kuwait and fears of a large-scale disruption, the oil price surged to $32 a barrel. However, oil prices retreated with the arrival of coalition troops and the start of the military campaign against Iraq. Although the battle settled the question of price, the issue of lost supplies required the cooperation of oil-producing countries, including Saudi Arabia. Oil supplies increased from several countries, including the United States, Venezuela, Mexico, and Norway. The largest increase, however, came from Saudi Arabia, raising output from 6.4 mbd in 1990 to 1998 mbd in 1991 (see table 5.13, Chapter 5). Moreover, Saudi Arabia's balancing effort included funding the war chest of the coalition. Without oil revenues, the kingdom could have hardly afforded to pay for the cost of driving Saddam Hussein out of Kuwait. In sum, the kingdom heavily relied on oil resources to balance with oil-consuming countries against Saddam Hussein's Iraq.

The Boom Era, 2000–2012

The strong demand from non-OECD countries helped lift up the oil market at the turn of the 21st century. Over the period 2001–2012, non-OECD countries experienced an 18 percent growth, reaching 43 mbd in 2012. On the other hand, OECD countries experienced stagnation or decline in demand. After a temporary increase from 49 mbd in 2001 to 2050 mbd in 2004, consumption declined to 46 mbd in 2012. On the production side, supply did not keep up with demand due to several shortages, namely, the strike of Petróleos de Venezuela SA, the US-led invasion of Iraq, and hurricanes Katrina and Rita. These shortages were partly compensated by higher production from Saudi Arabia. In 2002–2003, Saudi Arabia increased production by 1.3 mbd. Production continued to increase, reaching 9.3 mbd in 2005, and remained above 8 mbd, reaching 9.7 mbd in 2012.

Throughout this period, consuming countries were asymmetrically dependent on producers. Higher prices and consumption enriched the coffers of the kingdom. Oil revenue jumped from 183 trillion SR in 2002 to 2983 trillion SR in 2008. After a brief dip in 2009, oil revenues recovered in 2010 and reached 1,144 trillion SR in 2012. Awash with cash, the kingdom increased spending on human, social, and economic development as well as infrastructure. Spending on subsidies for food and social security also increased. Comparably, spending significantly increased on internal and external security. Allocations to security went up from 52 million SR in 2001 to 2120 million SR in 2012. Except for 2002 and 2009, when prices were relatively low, military spending continued to increase yearly, on average, 12 percent.

The vast supply of oil and wealth allowed the Saudi regime to lavishly spend on strategies of internal validation as well as internal and external balancing as a response to security challenges. In the case of Iran, Saudi Arabia became troubled with Iran's activities in the Levant, especially after the election of Mahmoud Ahmadinejad in 2005. Saudi Arabia used its oil revenues to counterbalance the Iranian influence in Iraq, Lebanon, and Palestine. At the same time, Saudi Arabia exploited its massive supply to pressure oil price downward in 2008. Oil price rose to a record high of $131.22 a barrel in July 2008, allowing Iran to amass a huge amount of oil revenues. To deflate the Iranian budget, Saudi Arabia cranked up oil production throughout 2008 until, finally, oil prices collapsed. In December 2008, oil price fell to $38.60 per barrel.

A massive oil budget allowed Saudi Arabia to drastically spend on internal validation and external balancing in response to the Arab Spring. In Egypt, the Saudi government aligned with the Egyptian military against the Muslim Brotherhood. Saudi money flowed into Egyptian government coffers after ousting President Mohammed Morsi. Similarly, Saudi Arabia provided significant financial aid to Yemen and Bahrain. At home, the Saudi government pledged to spend $66.7 billion on housing and $500 million on medical facilities. Other spending was allocated to raising minimum wages, unemployment allowance, and bonuses worth two months' salary. Moreover, all university students were promised a $500 allowance. In sum, the boom era provided the kingdom with the financial clout to buy off support inside and outside the kingdom.

What Lies Ahead for Saudi Arabia?

Although it is widely speculated that the post-petroleum order is approaching, most predictions show that oil will continue to dominate energy resources for the next 20 years. The figures in table 7.1 demonstrate that future demand for

TABLE 7.1 *World supply of primary energy, 2006–2030*

Energy type	Million tons		Growth (percentage)	Fuel shares (percentage)	
	2006	2030	2006–2030	2006	2030
Oil	4,031	5,360	1.2	37.3	32.7
Coal	2,989	4,655	1.9	27.6	28.4
Gas	2,400	3,993	2.1	22.2	24.4
Nuclear	731	1,022	1.4	6.8	6.2
Hydro	251	427	2.2	2.3	2.6
Biomass	349	674	2.8	3.2	4.1
Other renewable	61	258	6.2	0.6	1.6
Total	10,813	16,389	1.7	100	100

SOURCE: OPEC, WORLD OIL OUTLOOK 2008.

energy in general, including oil, is going up, not down. Even though the percentage growth rate of oil will be outpaced by renewable and other sources, oil will continue to claim the lion's share of total fuel consumption. The reason for oil's domination partly has to do with the cost of alternative sources of energy in comparison with oil prices, but also because switching from oil to other sources requires building up new infrastructures, which necessitates a huge amount of investment and time. Moreover, in many of the less developed countries, the lack of technology and financial means will also hinder diversification efforts.

Overall, world demands for energy will steadily climb, because of population growth and economic productivity. The world population is expected to increase from a historical level of 6.5 billion in 2005 to more than 8.3 billion by the year 2030. The largest increase will take place in developing countries (about 85 percent of the total population growth), which is also where oil demand is expected to rise significantly.[6] Economic growth plays a key factor in strengthening demand for oil. Because energy is the catalyst behind industrialization, transportation, and manufacturing activities, it is an essential prerequisite for economic development and growth. Most of the projected growth

6 UN, the Department of Economic and Social Affairs, "World Urbanization Prospects: The 2005 Revision." Accessed November 2014. http://www.un.org/esa/population/publications/WUP2005/2005WUP_FS4.pdf.

is expected to originate from developing countries in Asia, mainly China and India. Whereas China is projected to grow in the 2005–2030 period at a brisk average annual rate of 6.4 percent, India is not far behind with an average annual growth of 5.8 percent. During the same period, the annual rate of growth for the rest of the Asian countries is expected to reach an average of 4.6 percent. In contrast, the industrialized world's growth is expected to slow down. The United States is forecasted to grow at 2.6 percent, and OECD Europe at 2.3 percent. These projections imply that in 2030 China will have a GDP of $36 trillion, the highest in the world, accounting for 23 percent of the world gross domestic product (GDP). Second to China will be the United States with a GDP of $20.2 trillion, followed by OECD Europe with a GDP of $20 trillion; and in fourth place will be India with a GDP of $16.5 trillion.[7]

The implication of these economic projections compounded with population growth is that world demand for energy, including oil, will also steadily rise. World oil demand from 2006 to 2030 will rise by 29 mbd. The largest share of this rise is attributed to China and other Asian developing countries. By 2030 China is projected to have doubled its daily consumption, reaching 15.4 mbd, while South Asian countries, dominated by India, will have quadrupled their daily consumption, reaching 8.5 mbd. In contrast, the industrialized Western world is expected to experience lower growth for oil, because of technological innovations and governmental policies of environmental sustainability. In absolute terms, however, North America and Western Europe will remain the largest oil consumers within the next 20 years. From 2006 to 2030, in North America consumption is expected to increase from 25.3 mbd to 27.3 mbd, while in Western Europe consumption is expected to modestly increase from 15.7 mbd to 16.2 mbd.[8] In comparison with the developing world, OECD countries are projected to use five times more oil per person by 2030.

Meeting future oil demands will not only challenge oil consumers, but also oil producers. By 2030 many oil reserves around the world will be depleted and many oil producers will no longer have the capacity to produce for exports, except for a few countries. Six oil producers will dominate the oil market: Russia, Saudi Arabia, Iraq, Iran, Kuwait, and United Arab Emirates (see table 7.2).

Among the six countries, Saudi Arabia will remain the biggest producer with the largest proven reserves. As a result, neo-rentier theory predicts that Saudi Arabia is likely to employ strategies of high internal validation and internal balancing. At the same time, a tight market will create opportunities for the

7 GDP projections are in Purchasing Power Parity. See EIA, "International Energy Outlook 2008." Accessed February 2009. http://www.eia.doe.gov/oiaf/ieo/.

8 OPEC, "World Oil Outlook 2008," 32–36. Accessed February 2009. http://www.opec.org.

TABLE 7.2 *Saudi Arabia conventional liquids production,* 2005–2030*

	History (estimates)		Projections			
Regional/Country	**2005**	**2006**	**2015**	**2020**	**2025**	**2030**
Saudi Arabia (Million barrels per day)	11	11	12	13	13	14
Total world (Million barrels per day)	82	82	89	94	98	103
OPEC share of world production (percentage)	43	43	44	46	46	46
Persian Gulf share of world produciton (percentage)	29	29	29	30	31	31
Saudi Arabia share of world production (percentage)	13	13	13	13	13	13

SOURCE: EIA, INTERNATIONAL ENERGY OUTLOOK 2008, WORLD CONVENTIONAL LIQUIDS PRODUCTION BY REGION AND COUNTRY, REFERENCE CASE.
* EIA defines conventional liquid as "crude oil and lease condensate, natural gas liquids, and refinery gain." See EIA, International Energy Outlook 2009: World Conventional Liquids Production by Region and Country, Reference Case.

kingdom to employ external validation and external balancing. Such predictions, of course, must be taken with reservations, for the complexity involving social science events are far too great to make forecasts about their future with a definitive authority. Nevertheless, such predictions provide a glimpse of Saudi Arabia's security behavior in the coming oil world order.

Bibliography

Aarts, Paul and Gerd Nonneman. eds., *Saudi Arabia in the Balance: Political Economy, Society, Foreign Affairs*. London: Hurst & Company, 2005.

Abir, Mordechai. *Saudi Arabia: Government, Society, and the Gulf Crisis*. London: Routledge, 1993.

Ajami, Fouad, "The End of Pan-Arabism." *Foreign Affairs,* 57 (Winter 1978/1979): 355–373.

Al-Rasheed, Madawi. "Saudi Internal Dilemmas and Regional Reponses to the Arab Uprisings." In *The New Middle East: The Protest and Revolution in the Arab World*, ed. Fawaz Gerges. Cambridge University Press, 2014.

Anderson, Irvine H., Aramco. *The United States, and Saudi Arabia: A Study of the Dynamics of Foreign Oil Policy, 1933–1950*. Princeton: Princeton University Press, 1981.

Anderson, Lisa. "The State in the Middle East and North Africa." *Comparative Politics* 20 (October 1987): 1–18.

Anderson, Lisa. "Lawless Government and Illegal Opposition: Reflection on the Middle East." *Journal of International Affairs* 40 (spring 1987): 219–232.

Anderson, Lisa. *The State and Social Transformation in Tunisia and Libya, 1880–1980*. Princeton: Princeton University Press, 1986.

Art, Robert J. and Kenneth N. Waltz. *The Use of Force: Military Power and International Politics*. Rowman & Littlefield Publishers, Inc., 2003.

Askari, Hossein. *Saudi Arabia's Economy: Oil and the Search for Economic Development.* London: Jai Press, In., 1990.

Ayoob, Mohammed. "Inequality and Theorizing in International Relations: The Case for Subaltern Realism." *International Studies Review* 4 (Autumn 2002): 27–48.

Ayoob, Mohammed. "Review: The Security Problematic of the Third World." *World Politics* 43 (January 1991): 257–283.

Ayoob, Mohammed. "The Third World in the System of States: Acute Schizophrenia or Growing Pains?" *International Studies Quarterly* 33 (March 1989): 67–79.

Baldwin, David A. "Interdependence and Power: A Conceptual Analysis." *International Organization* 34 (Autumn 1980): 471–506.

Baldwin, David A. "Power Analysis and World Politics: New Trends versus Old Tendencies." *World Politics* 31(January 1979): 161–194.

Bamberg, James. *British Petroleum and Global Oil, 1950–1975: The Challenge of Nationalism*. Cambridge: Cambridge University Press, 2000.

Beaumont, Peter. "Water and Development in Saudi Arabia." *The Geographical Journal* 143 (March 1977): 42–60.

Beblawi, Hazem and Giacomo Luciani, eds. *The Rentier State*. New York: Croom Helm, 1987.

Beetham, David. *The Legitimation of Power*. London: Macmillan Education Ltd, 1991.

Bendix, Reinhard. *Kings or People: Power and the Mandate to Rule.* Berkeley: University of California Press, 1974.

Blanchard, Christopher M. "Saudi Arabia: Background and U.S. Relations." *CRS Report for Congress*. Accessed March 2008. http://www.fas.org.

Bronson, Rachel. *Thicker Than Oil: America's Uneasy Partnership with Saudi Arabia.* Oxford: Oxford University Press, 2006.

British Petroleum. *Statistical Review of World Energy June 2008.* Accessed August 2008. http://www.bp.com.

Bueno de Mesquita, Bruce, James D. Morrow, Randolph M. Siverson, and Alastair Smith. "An Institutional Explanation of the Democratic Peace." *The American Political Science Review* 93 (December 1999): 791–807.

Buzan, Barry. "Economic Structure and International Security: The Limits of the Liberal Case." *International Organization* 38 (Autumn 1984): 597–624.

Carapico, Sheila. "No Exit: Yemen's Existential Crisis." In *ArabThe Revolts: Dispatches on Militant Democracy in the Middle East*, eds. David McMurray and Amanda Ufheil-Somers. Bloomington: India University Press, 2013.

Chatelus, Michel and Yves Schemeil "Towards a New Political Economy of State Industrialization in the Arab Middle East." *International Journal of Middle East Studies* 16 (May 1984): 251–265.

Chaudhry, Kiren Aziz. *The Price of Wealth: Economies and Institutions in the Middle East*. Ithaca: Cornell University Press, 1997.

Cook, M.A., ed. *Studies in the Economic History of the Middle East.* Oxford: Oxford University Press, 1970.

Cooper, Richard N. "Economic Interdependence and Foreign Policy in the Seventies." *World Politics* 24 (January 1972): 159–181.

Copeland, Dale. "Economic Interdependence and the Outbreak of War." paper presented to University of Virginia Department of Government's faculty workshop (March 1995).

Cordesman, Anthony. *Saudi Arabia Enters the Twenty-First Century: The Political, Foreign Policy, Economic, and Energy Dimension*. Praeger Publishers: Westport, 2003), 44–51.

Cordsman, Anthony H. and Nawaf E. Obaid. *National Security in Saudi Arabia: Threats, Responses, and Challenges.* Washington D.C: Center for Strategic and International Studies, 2005.

Cordesman, Anthony H. and Abraham R. Wagner, *the Lessons of Modern War: The Iran-Iraq War*, Volume II. Colorado: Westview Press, 1990.

Crystal, Jill. "Coalitions in Oil Monarchies: Kuwait and Qatar." *Comparative Politics* 21, (July 1989): 427–443.

Dahl, Robert A. "The Concept of Power." *Behavioral Science* 2 (July 1957): 201–215.

Daoudi M.S., and M.S. Dajani. "The 1967 Oil Embargo Revisited." *Journal of Palestine Studies*, 13 (Winter 1984).

David, Steven R. "Why the Third World Still Matters." *International Security* 17 (Summer 1992): 50–85.

David, Steven R. "Explaining Third World Alignment." *World Politics* 43 (January 1991): 233–256.

Delacroix, Jacques. "The Distributive State in the World System." *Studies in Comparative International Development* 15 (1980): 3–21.

Dawisha, Adeed and I. William Zartman eds. *Beyond Coercion: The Durability of the Arab State*. New York: Croom Helm, 1988.

Dawisha, Adeed. *Saudi Arabia's Search for Security*, Adelphi Papers 158. London: International Institute for Strategic Studies, 1979.

Doyle, Michael. *Ways of War and Peace*. New York: W.W. Norton & Company, Inc., 1997.

Esposito, John L., ed. *The Iranian Revolution and Its Global Impact*. Miami: Florida International University Press, 1990.

Evans, Peter, Dietrich Rueschemeyer and Theda Skocpol, eds. *Bringing the State Back In*. Cambridge: Cambridge University Press, 1985.

Evera, Stephen W. Van. *Causes of War Power and the Roots of Conflict*. Ithaca: Cornell University Press, 1999.

Facwcett, Louise, ed. *International Relations of the Middle East*. New York: Oxford University Press, 2005.

Fandy, Mamoun. *Saudi Arabia and the Politics of Dissent*. London: Macmillan Press Ltd, 1999.

Fandy, Mamoun. "Serious Threats to Saudi Arabia's Stability." *Christina Science Monitor* 88 (January 1996).

Fearon, James D. "Domestic Politics, Foreign Policy, and Theories of International Relations." *Annual Review of Political Science* 1 (1998): 139–165.

Gaury, Gerald de. *Faisal: King of Saudi Arabia*, Appendix I. London: Arthur Barker Limited, 1966.

Gause, F. III Gregory. "Saudi Arabia Over a Barrel." *Foreign Affairs* 79 (May/June 2000): 84–94.

Gause, III, F. Gregory. *Political Opposition in the Gulf Monarchies*. Florence: European University Institute Working Papers, 2000.

Gause, III, F. Gregory. *Oil Monarchies*. New York: Council on Foreign Relations Press, 1994.

Gerges, Fawaz A. "The Kennedy Administration And the Egyptian-Saud Conflict in Yemen: Co-opting Arab Nationalism." *Middle East Journal*, Volume 49, No. 2 (Spring 1995): 292–311.

Gilpin, Robert. *The Political Economy of International Relations*. New Jersey: Princeton University press, 1987.

Greenstein, F.I. and N.W. Polsby, eds. *Handbook of Political Science*, Volume II Reading: Addison-Wesley, 1975.

Grimmett, Richard F. "Conventional Arms Transfers to Developing Nations, 1993 to 2000." *Mediterranean Quarterly* 13 (Spring 2002): 36–55.

Habib, John S. *Ibn Sa'ud's Warriors of Islam, the Ikhwan of Najd and Their Role in the Creation of the Saudi Kingdom*. Leiden: E.J Brill, 1978.

Halliday, Fred. *Arabia without Sultans*. Reprint, 1975. Middlesex, England: Penguin Books, 1974.

Helms, Christine Moss. *The Cohesion of Saudi Arabia: Evolution of Political Identity*. London: Croom Helm, 1981.

Herb, Michael. *All in the Family: Absolutism, Revolution, and Democracy in the Middle Eastern Monarchies*. New York: State University of New York Press, 1999.

Hinnebusch, Raymond and Anoushiravan Ehteshami, eds. *The Foreign Policies of Middle East States*. Colorado: Lynne Rienner, 2002.

Hirschman, Albert O. *National Power and the Structure of Foreign Trade*. expanded edition, 1980, Berkeley: University of California Press, 1945.

Hobbes, Thomas. *Leviathan*. London: Penguin Books, reprinted 1985.

Holden, David and Richard Johns. *The House of Saud*. London: Pan Books Ltd, 1982.

Hudson, Michael C. *Arab Politics: The Search for Legitimacy*. New Haven: Yale University Press, 1980.

Hudson, Michael. "Public Opinion, Foreign Policy, and the Crisis of Legitimacy in Arab Politics." *Journal of Arab Affairs* 5 (1986).

Humphreys, Macartan, Jeffrey D. Sachs, and Joseph E. Stiglitz. *Escaping the Resource Curse*. New York: Columbia University Press, 2007.

Hunter, Shireen T. *Iran and the World: Continuity in a Revolutionary Decade*. Bloomington: Indian University Press, 1990.

Ikenberry, John G. "The Irony of State Strength: Comparative Responses to the Oil Shocks in the 1970s." *International Organization* 40 (Winder, 1986): 105–137.

International Crisis Group. "Saudi Arabia Backgrounder: Who are the Islamists?" Middle East Report Number 31, 21 (September 2004).

International Monetary Fund (IMF). International Financial Statistics Yearbook. Accessed August 2007. http://www.imf.org.

Ion, D.C., *Availability of World Energy Sources*. ondon: Graham and Trotman, 1980.

Jankowski, P. James. *Nassers' Egypt, Arab Nationalism, and the United Arab Republic*. Boulder: Lynne Publishers, 2001.

Jervis, Robert. "Cooperation under the Security Dilemma." *World Politics* 30 (January 1978): 167–214.

Kennedy, Paul. *The Rise and Fall of the Great Powers: Economic Change and Military Conflict from 1500 to 2000*. New York: Random House, 1987.

Keohane, Robert O. and Joseph S. Nye. *Power and Interdependence*: World Politics in Transition, 3rd ed. New York: Longman, 2001.

Keohane, Robert O. and Joseph S. Nye. "Power and Interdependence." *International Organizations* 41 (Autumn 1987): 513–530.

Keohane, Robert O. *After Hegemony: Cooperation and Discord in the World Political Economy*. Princeton: Princeton University Press, 1984.

Keohane, Robert O. ed. *Neorealism and Its Critics*. New York: Cambridge University Press, 1986.

Kerr, Cortni and Toby Jones. "A Revolution Paused in Bahrain." In *ArabThe Revolts: Dispatches on Militant Democracy in the Middle East*, eds. David McMurray and Amanda Ufheil-Somers. Bloomington: India University Press, 2013.

Kerr, Malcolm H. and El Sayed Yassin eds. *Rich and Poor States in the Middle East: Egypt and the New Arab Order.* Colorado: Westview Press, Inc.

Khan, M.A. Muqtedar. *Jihad for Jerusalem.* Greenwood Publishing Group, 2004.

The Kingdom of Norway, Statistics Norway (Statistisk sentralbyrå). *Historical Statistics 1994*. Accessed June 2009. http://www.ssb.no.

Kingdom of Saudi Arabia, Ministry of Economy and Planning. *Annual Statistical Books 1965,1970,1975,1980,1985,1990,1995, 1999*. Accessed March 2007. http://www.planning .gov.sa.

Kingdom of Saudi Arabia, Ministry of Economy and Planning. *The First Development Plan, 1970–1975; Second Development Plan, 1975–1980; Third Development Plan, 1980–1985; Fourth Development Plan, 1985–1990; Fifth Development Plan, 1990–1995; Sixth Development Plan, 1995–2000; Seventh Development Plan, 2000–2005; Eighth Development Plan, 2000–2005*. Accessed July 2007. http://www.planning.gov.sa.

Kingdom of Saudi Arabia, Saudi Arabia Monetary Agency (SAMA). *Annual Statistical Bulletin.* Accessed March 2007. http://www.sama.gov.sa.

Kissinger, Henry. *Years of Upheaval*. Boston: Little, Brown and Company, 1982.

Kirabaeva, K., F. Parodi, and Y. Said. "Iraq Selected Issues." (April 26, 2013). IMF Country Report No. 13/218. Accessed August 2014. http://www.imf.org/external/pubs/ft/scr/2013/cr13218.pdf.

Korany, Bahgat and Ali E. Hillal Dessouki, eds. *The Foreign Policies of Arab States*. Colorado: Westview Press, 1984.

Krimly, Rayed Khalid. *The Political Economy of the Rentier States: A Case Study of Saudi Arabia*. PhD diss., George Washington University, 1993.

Kumins, Lawrence and Robert Bamberger. "Oil and Gas Disruption From Hurricanes Katrina and Rita." CRS Report for Congress, RL 33124. Accessed September 2014. http://fpc.state.gov/documents/organization/55824.pdf.

Lake, David A. "Powerful Pacifists: Democratic States and War." *American Political Science Review* 86 (March 1992): 24–37.

Lamborn, Alan C. "Power and Politics of Extraction." *International Studies Quarterly* 27(June 1983): 125–146.

Lauren, Paul Gordon ed. *Diplomacy: New Approaches in History, Theory, and Policy.* New York: The Free Press, 1979.

Lesch, Ann Mosely. "Contrasting Reactions to the Persian Gulf Crisis: Egypt, Syria, Jordan, and the Palestinians." *Middle East Journal* 45 (Winter 1991): 30–50.

Levy, Jack S. "The Offensive/Defensive Balance of Military Technology: A Theoretical and Historical Analysis." *International Studies Quarterly* 28 (January 1984): 219–238.

Long, David E. *The Kingdom of Saudi Arabia.* Florida: University Press of Florida, 1997.

Machiavelli, Niccoló. *The Prince,* 2nd ed. Chicago: The University of Chicago Press, 1998.

Mackey, Sandra. *The Iranians: Persian, Islam and the Soul of a Nation.* New York: Penguin Group, 1996.

Mastanduno, Michael. "Economic Statecraft, Interdependence, and national security: Agenda for research." *Security Studies* 9 (September 1999): 288–316.

Mastanduno, Michael. "Economics and Security in Statecraft and Scholarship." *International Organization,* 52, 4 (Autumn, 1998): 828–854.

Mastanduno, Michael, David A. Lake, and G. John Ikenberry. "Toward a Realist Theory of State Action." *International Studies Quarterly* 33 (December, 1989): 457–474.

Mearsheimer, John J. *The Tragedy of Great Power Politics.* New York: W.W. Norton Company, 2001.

Migdal, Joel S. *Strong Societies and Weak States.* Princeton: Princeton University Press, 1988.

Milani, Mohsen M. "Iran's Strategies and Objectives in Post-Saddam Iraq." In *Iraq, Its Neighbors, and the United States: Competition, Crisis, and the Reordering of Power,* ed.

Henri J. Barkey, Scott B. Lasensky, and Phebe Marr. Washington DC: United States Institute of Peace, 2011.

Mill, John Stuart. *Principles of Political Economy.* 4th reprint. Clifton: Kelley, 1973.

Moravcsik, Andrew. "Taking Preferences Seriously: A Liberal Theory of International Politics." *International Organization* 51 (Autumn, 1997): 513–553.

Morgenthau, Hans J. *Politics Among Nations: The Struggle for Power and Peace,* 5th ed. New York: Alfred A. Knopf, 1973.

Most, Benjamin A. and Harvey Starr. "International Relations, Foreign Policy Substitutability, and 'Nice' Laws." *World Politics* 36 (April 1984): 383–406.

Muller, John. *Retreat from Doomsday.* New York: Basic Books, 1989.

Murden, Simon W. *Emergent Regional Powers and International Relations in the Gulf: 1988–1991.* Reading: Ithaca Press, 1995.

Myrdal, Gunnar. *Asian Drama: An Inquiry into the Poverty of Nations,* vol. 2. New York: 1968.

Niblock, Tim, ed. *State, Society and Economy in Saudi Arabia.* London: Croom Helm Ltd, 1983.

Niblock, Tim and Monica Malik. *The Political Economy of Saudi Arabia*. London: Routledge, 2007.

Organski, A.F.K. *World Politics*, 2nd edition. New York: Alfred Al. Knopf, 1968.

Parra, Francisco. *Oil Politics: A Modern History of Petroleum*. London: I.B. Tauris & Co. Ltd, 2004.

Philby, Harry St. John. *Arabian Days: An Autobiography*. London: Robert Hale Limited, 1948.

Philby, Harry St. John. *Saudi Arabia*. New York: Arno Press, 1972.

Philby, Harry St. John. *The Heart of Arabia: A Record of Travel and Exploration*, volume II. London: Constable & Company, 1922.

Organization of Petroleum Exporting Countries (OPEC). "Annual Statistical Bulletin 2004." Accessed April 2008. http://www.opec.org/library

Organization of Petroleum Exporting Countries (OPEC). "Annual Statistical Bulletin 2005." Accessed December 2006. http://www.opec.org/library.

Organization of Petroleum Exporting Countries (OPEC). "Annual Statistical Bulletin 2006." Accessed August 2008. http://www.opec.org/library.

Organization of Petroleum Exporting Countries (OPEC). "Annual Statistical Bulletin 2007." Accessed January 2008. http://www.opec.org/library.

Organization of Petroleum Exporting Countries (OPEC). "Oil Outlook 2008." Accessed February 2009. http://www.opec.org.

Organization of Petroleum Exporting Countries (OPEC). "Annual Statistical Bulletin 2012." Accessed August 2014. http://www.opec.org/library.

Quandt, William B. *Decade of Decisions*. California: California University Press, 1977.

Quandt, William B. *Saudi Arabia in the 1980s: Foreign Policy, Security, and Oil*. Washington, D.C.: The Brookings Institution, 1981.

Quandt, William B. *DavidCamp: Peace Making and Politics*. Washington D.C.: The Brookings Institution, 1986.

Ray, James Lee. "Does Democracy Cause Peace?" *Annual Review of Political Science*, 1 (1998): 27–46.

Al-Rasheed, Madawi. *A History of Saudi Arabia*. London: Cambridge University Press, 2002.

Richards, Alan and John Waterbury. *A Political Economy of the Middle East*. Boulder: Westview Press, 1996.

Rihani, Ameen. *Ibn Sa'ud of Arabia: His People and His Land*. London: Constable, 1928.

Rose, Gideon. "Neoclassical Realism and Theories of Foreign Policy." *World Politics* 51, (October 1998): 144–172.

Ross, Michael L. "Review: The Political Economy of the Resource Curse." *World Politics* 51 (January 1999): 297–322.

Sachs, Jeffrey D. and Warner M. Andrew. "The Curse of Natural Resources." *European Economic Review* 45 (2001): 827–838.

El-Sadat, Anwar. *In Search of Identity*. London: Collins, 1978.

Safran, Nadav. *Saudi Arabia: The Ceaseless Quest for Security*. Ithaca: Cornell University Press, 1988.

Salame, Ghassan. "Political Power and the Saudi State." *MERIP Reports* (October 1980): 5–22.

Sampson, Anthony. *The Seven Sisters: The Great Oil Companies and the World They Made*. New York: The Viking Press, 1975.

Schofield, Daniel, ed. *The Kingdom of Saudi Arabia*. London: Stacey International, 1986.

Schumpeter, Joseph. *Imperialism and Social Classes*, trans., Heinz Norden. Oxford: Blackwell, 1951.

Seymour, Ian. *OPEC Instrument of Change*. London: The Macmillan Press LTD, 1980.

Seznec, Jean-Francosis. "Democratization in the Arab World? Stirrings in Saudi Arabia." *Journal of Democracy* 13 (October 2002), 33–40.

Sharabi, Hisham, ed. *Theory, Political and the Arab World: Critical Responses*. Routledge: New York, 1990.

Shojai, Siamack, ed. *The New Global Oil Market: Understanding Energy Issues in the World Economy*. Westport, CT: Praeger, 1995.

Shojai, Siamack and Bernard S. Katz, ed. *The Oil Market in the 1980's: A Decline of Decline*. New York: Praeger Publishers, 1992.

Smith, Adam. *The Wealth of Nations*. Edited by Edwin Cannan. New York: The Modern Library, 1994.

Stinchcombe, Arthur L. *Constructing Social Theories*. New York: Harcourt, Brace & World, Inc., 1968.

Stockholm International Peace Research Institute (SIPRI). "Trends in World Military Expenditures, 2013." Accessed August 2014. http://books.sipri.org/product_info?c_product_id=476.

Tessler, Mark. *Area Studies and Social Science: Strategies for Understanding Middle East Politics*. Bloomington: Indiana University Press, 1999.

Tilly, Charles, ed. *The Formation of National States in Western Europe*. Princeton: Princeton University Press, 1975.

United Nations, Department of Economic and Social Affairs, Population Division. "The World population Prospects: The 2008 Revision." Accessed February 2009. http://www.un.org/esa/desa.

United Nations, Department of Economic and Social Affairs, Population Division. "World Urbanization Prospects: The 2005 Revision." Accessed November 2014. http://www.un.org/esa/population/publications/WUP2005/2005WUP_FS4.pdf

United States Training Mission (USMTM) to the Kingdom of Saudi Arabia. Accessed April 2008. http://usmtm.org/index.html.

United States Energy Information Administration (EIA). "Annual Energy Review 2006." Report no. DOE/IEA-0384. Accessed 2008. http://www.eia.gov/totalenergy/data/annual/previous.cfm#2006.

United States Energy Information Administration (EIA). "Annual Energy Review, 2007." Accessed October 2007. http://www.eia.doe.gov/emeu/aer/petro.html.

United States Energy Information Administration (EIA). "International Energy Outlook 2008." Accessed February 2009. http://www.eia.doe.gov/oiaf/ieo/.

United States Energy Information Administration (EIA). "International Energy Outlook 2009." Accessed February 2009. http://www.eia.doe.gov/oiaf/ieo/.

United States Energy Information Administration (EIA). "STEO Supplement: Why are oil prices so high?" Accessed July 2014. http://www.eia.gov/forecasts/steo/special/pdf/high-oil-price.pdf.

Verba, Sidney. "Some Dilemmas in Comparative Research." *World Politics* 20 (October 1967): 111–127.

Vitalis, Robert. "The Closing of the Arabian Oil Frontier." *Middle East Report* 204 (Fall 1997): 15–21.

Walt, Stephen M. *The Origins of Alliances*. Ithaca: Cornel University Press, 1990.

Waltz, Kenneth N. *Theory of International Politics*. San Francisco: McGraw-Hill, Inc., 1979.

Waterbury, John. "The 'Soft State' and the Open Door: Egypt's Experience with Economic Liberalization, 1974–1984." *Comparative Politics*, 18 (October 1985).

Weber, Max. *The Theory of Social and Economic Organization*. Translated by A.M. Henderson and Talcott Parsons. New York: The Free Press, 1964.

Wehrey, Frederic, Theodore W. Karasik, Alireza Nader, Jeremy Ghez, Lydia Hansell, and Robert A. Guffey. *Saudi-Iranian Relations Since the Fall of Saddam: Rivalry, Cooperation, and Implications for U.S. Policy*. Rand National Security Research Division, 2009. Accessed August 2014. http://www.rand.org/pubs/monographs/MG840.html.

Wohlforth, William Curti. *The Elusive Balance: Power and Perceptions during the Cold War*. Ithaca, N.Y.: Cornell University Press, 1993.

Yergin, Daniel. *The Prize: The Epic Quest for Oil, Money, and Power*. New York: Simon & Schuster, 1991.

Young, Arthur N. *Saudi Arabia: The Making of a Financial Giant*. New York: New York University Press, 1983.

Zakaria, Fareed. *From Wealth to Power: the Unusual Origins of America's World Role*. Princeton: Princeton University Press, 1998.

News Media

Biddle, Fred Marc. "The Saudi Strategy in Cutting Oil Prices," *Chicago Tribune*, September 22, 1985.

Black, Ian. "Saudi Arabian Security Forces Quell 'Day of Rage' Protests," *Guardian*, March 11, 2011.

Fattah, Hassan M. "Bickering Saudis Struggle for an Answer to Iran's Rising Influence in the Middle East," *New York Times*, December 22, 2006.

Hooper, John. "Don't Rock a Boat Making Headway: The Latest Opec Meeting," *Guardian*, June 26, 1987.

Ibrahim, Youssef M. "Confrontation in the Gulf Foreigners in Iraq will Share in Hardships; Baghdad Warns; Saudis to Increase Oil Output; Challenge to OPEC," *New York Times*, August 19, 1990.

Ibrahim, Youssef M. "Saudi Arabia Cracks Down on Islamic Militants, Seizing Many," *New York Times*, September 22, 1994.

Ibrahim, Youssef M. "Saudi Arabia Arrests 110 Muslim Militants," *New York Times*, September 28, 1994.

Ibrahim, Youssef M. "New Pipelines Are Reducing Persian Gulf's Strategic Role," *New York Times*, October 07, 1987.

Ibrahim, Youssef M. "Saudi King Ousts 7 Senior Clerics for Acts of Criticism by Omission," *New York Times*, December 15, 1992.

Kraus, James R. "U.S. Banks See Borrowing Rise in Persian Gulf," *The American Banker*, March 5, 1998.

Nordland, Rod. "Saudi Arabia Promises to Aid Egypt's Regime," *New York Times*, August 19, 2013.

Ottaway, David B. "Persian Gulf War Posing New Threat to U.S. Interests; Iraq Targets Iran's Main Oil Terminal," *Washington Post*, September 27, 1985.

Romero, Simon and Michael Slackman, and Clifford J. Levy. "Oil-Rich Countries Face a Reckoning," *New York Times*, October 20, 2008.

Sanger, David E. "The Price Confronting Iraq in the Oil Market; Japanese Joining in Embargo Despite Impact on Economy," *New York Times*, August 06, 1990.

Sciolino, Elaine, Jeff Gerth and Douglas Jehl. "Saudi Kingdom Shows Cracks U.S. Aides Fear," *New York Times*, June 30, 1996.

Stout, David. "U.S., Citing Security, Shuts Saudi Embassy Briefly," *New York Times*, October 7, 1998.

Tagliabue, John. "An OPEC Meeting Seems at Impasse," *New York Times*, August 03, 1986.

Tisdall, Simon. "WikiLeaks Cables: Saudi Arabia Rated a Bigger Threat to Iraqi Stability than Iran," *Guardian*, December 5, 2010.

Young, Karen De. "British Get $4.5 Billion Plane Order; Saudi Purchase Had Green Light from White House," *Washington Post*, September 27, 1985.

"Americans Given Option of Leaving Saudi Arabia," *New York Times*, July 23, 1996.

"Saudi Monarch Announces Billions in Handouts." Accessed July 2014. http://www.aljazeera.com/news/middleeast/2011/03/2011318174117916648.html.

"The Oil Weapon," *Washington Post*, August 05, 1990.

"Unable to Balance the Books, *"The Economist*. Accessed August 2014. http://www.economist.com/blogs/pomegranate/2014/06/yemens-economy.

Index

www.ingramcontent.com/pod-product-compliance
Lightning Source LLC
Chambersburg PA
CBHW071350220926
51613CB00008B/31
9789004277748